Smart Cities and Innovative Urban Technologies

Over the past decade smart urban technologies have begun to blanket our cities, forming the backbone of a large, intelligent infrastructure. Along with this development, dissemination of the smart cities ideology has had a significant imprint on urban planning and development. *Smart Cities and Innovative Urban Technologies* focuses on the concepts of smart cities and innovative urban technologies. It contains research that provides insight into spatial formations of information and communication technologies, and knowledge production practices from various perspectives—including analyses of public and private sectors together with NGOs and other stakeholders. It provides a state-of-the-art analysis from a multidisciplinary point-of-view in urban studies.

Contributions in this edited volume include theoretical developments as well as empirical analyses. This book will be of great use to various audiences including academics as well as practitioners, spatial developers, planners and public administrators, in order to increase understanding of the dynamics and factors affecting smart cities' conceptual maturation and their physical emergence. Information generated in these chapters, particularly regarding the challenges and obstacles of smart cities and innovative urban technologies, are intended to be of benefit to the key local actors making decisions in their cities and/or peripheral locations.

This book was originally published as a special issue of the *Journal of Urban Technology*.

Tommi Inkinen is Professor and Research Director at the University of Turku, Finland. He is the current Chairman of the International Geographical Union's (IGU) Innovation, Information and Technology Commission and serves on the editorial boards of several international journals.

Tan Yigitcanlar is Associate Professor in the School of Built Environment, Queensland University of Technology, Australia. He also carries out an Honorary Professor role in the School of Technology, Federal University of Santa Catarina, Florianopolis, Brazil, and Founding Director positions at Urban Studies Lab, Australia, and the Australia–Brazil Smart City Research and Practice Network.

Mark Wilson is Professor in the School of Planning, Design and Construction, Michigan State University, USA. He is Director of the Planning, Design and Construction doctorate programme. His research and teaching interests address urban planning, information technology, economic geography, public policy and non-profit organizations.

Smart Cities and Innovative Urban Technologies

Edited by
Tommi Inkinen, Tan Yigitcanlar and Mark Wilson

LONDON AND NEW YORK

First published 2021
by Routledge
2 Park Square, Milton Park, Abingdon, Oxon OX14 4RN

and by Routledge
52 Vanderbilt Avenue, New York, NY 10017

Routledge is an imprint of the Taylor & Francis Group, an informa business

British Library Cataloguing in Publication Data
A catalogue record for this book is available from the British Library

ISBN 13: 978-0-367-67793-0

Typeset in MinionPro
by Newgen Publishing UK

Publisher's Note
The publisher accepts responsibility for any inconsistencies that may have arisen during the conversion of this book from journal articles to book chapters, namely the inclusion of journal terminology.

Disclaimer
Every effort has been made to contact copyright holders for their permission to reprint material in this book. The publishers would be grateful to hear from any copyright holder who is not here acknowledged and will undertake to rectify any errors or omissions in future editions of this book.

Contents

Citation Information

The chapters in this book were originally published in the *Journal of Urban Technology*, volume 26, issue 2 (April 2019). When citing this material, please use the original page numbering for each article, as follows:

Introduction
Smart Cities and Innovative Urban Technologies
Tommi Inkinen, Tan Yigitcanlar and Mark Wilson
Journal of Urban Technology, volume 26, issue 2 (April 2019), pp. 1–2

Chapter 1
Smart City Planning from an Evolutionary Perspective
N. Komninos, C. Kakderi, A. Panori and P. Tsarchopoulos
Journal of Urban Technology, volume 26, issue 2 (April 2019), pp. 3–20

Chapter 2
Smart Cities and Mobility: Does the Smartness of Australian Cities Lead to Sustainable Commuting Patterns?
Tan Yigitcanlar and Md. Kamruzzaman
Journal of Urban Technology, volume 26, issue 2 (April 2019), pp. 21–46

Chapter 3
The (In)Security of Smart Cities: Vulnerabilities, Risks, Mitigation, and Prevention
Rob Kitchin and Martin Dodge
Journal of Urban Technology, volume 26, issue 2 (April 2019), pp. 47–65

Chapter 4
E-Capital and Economic Growth in European Metropolitan Areas: Applying Social Media Messaging in Technology-Based Urban Analysis
Juho Kiuru and Tommi Inkinen
Journal of Urban Technology, volume 26, issue 2 (April 2019), pp. 67–88

Chapter 5
How to Overcome the Dichotomous Nature of Smart City Research: Proposed Methodology and Results of a Pilot Study
Luca Mora, Mark Deakin, Alasdair Reid and Margarita Angelidou
Journal of Urban Technology, volume 26, issue 2 (April 2019), pp. 89–128

Chapter 6

"Mapping" Smart Cities
Becky P. Y. Loo and Winnie S. M. Tang
Journal of Urban Technology, volume 26, issue 2 (April 2019), pp. 129–146

Chapter 7

Towards Post-Anthropocentric Cities: Reconceptualizing Smart Cities to Evade Urban Ecocide
Tan Yigitcanlar, Marcus Foth and Md. Kamruzzaman
Journal of Urban Technology, volume 26, issue 2 (April 2019), pp. 147–152

For any permission-related enquiries please visit:
www.tandfonline.com/page/help/permissions

Notes on Contributors

Margarita Angelidou is Senior Researcher at URENIO Research, Aristotle University of Thessaloniki, Greece.

Mark Deakin is Professor of Built Environment and Head of the Centre for Smart Cities at Edinburgh Napier University's School of Engineering and the Built Environment, UK.

Martin Dodge is Senior Lecturer in Human Geography at the University of Manchester, UK.

Marcus Foth is Professor of Urban Informatics in the QUT Design Lab, Queensland University of Technology, Brisbane, Australia.

Tommi Inkinen is Professor in the Department of Geosciences and Geography and the Research Director of the Centre for Maritime Studies at the University of Turku, Finland.

Christina Kakderi is a Regional Economist and Researcher focusing on systems of innovation and smart innovation environments (national and regional innovation systems, technology policy, social networks and intelligent cities/districts).

Md. Kamruzzaman is Associate Professor in the Faculty of Art, Design and Architecture, Monash University, Melbourne, Australia.

Rob Kitchin is Professor and ERC Advanced Investigator at Maynooth University, Ireland, and Managing Editor of the international journal *Dialogues in Human Geography*.

Juho Kiuru is a PhD student in the Department of Geosciences and Geography at the University of Helsinki, Finland.

Nicos Komninos is Professor in the Faculty of Engineering and Director of URENIO Research at the Aristotle University of Thessaloniki, Greece.

Becky P. Y. Loo is Professor and Head of the Department of Geography, The University of Hong Kong, HK.

Luca Mora is Lecturer in Urban Innovation Dynamics at the Business School of Edinburgh Napier University's School of Engineering and the Built Environment, UK.

Anastasia Panori is an Electrical and Computer Engineer. Since 2012, she has participated in various research projects funded by the European Union and is currently a Researcher at URENIO Research, Aristotle University of Thessaloniki, Greece.

Alasdair Reid is Research Fellow in the Centre for Smart Cities at Edinburgh Napier University's School of Engineering and the Built Environment, UK.

Winnie S. M. Tang is Founder and Chairman of Esri China (Hong Kong) and Adjunct Professor in the Department of Computer Science, The University of Hong Kong, HK.

P. Tsarchopoulos is Senior Researcher at URENIO Research, Aristotle University of Thessaloniki, Greece.

Mark Wilson is Professor in the School of Planning, Design and Construction, Michigan State University, USA.

Tan Yigitcanlar is Associate Professor in the School of Built Environment, Queensland University of Technology, Brisbane, Australia.

Smart Cities and Innovative Urban Technologies

Over the past two decades smart urban technologies have begun to form the backbone of new and large "intelligent" infrastructures, particularly in the privileged corners of the developed world—e.g., Amsterdam, Barcelona, Melbourne, San Francisco, and Singapore. Along with this development, dissemination of the smart city ideology has had a significant imprint on the planning and development of cities and their regions. Given the increasing popularity of the topic, this special issue concentrates on two highly interrelated concepts—*smart cities* and *innovative urban technologies*—and contains scholarly writings that provide some invaluable insights into the spatial effects of information and communication technologies and into knowledge production practices of our cities and societies.

This special issue consists of seven articles. These writings target the academic community as well as practitioners, such as spatial developers, planners, and public administrators, in order to increase the understanding of the dynamics and factors affecting the conceptual maturation of smart cities. Under the smart cities and intelligent urban technologies conceptual umbrella, the foci of these articles include economic growth, technological innovations, and social and political aspects of technological progression taking place in urban environments. The special issue also pays particular attention to global cities and relevant extensive analyses.

Information generated in these writings, particularly regarding the challenges and opportunities of smart cities and innovative urban technologies, are intended to benefit key local actors when they make decisions in their cities and/or peripheral locations. We believe this special issue will shed further light on the generation of a clearer understanding of smart cities, technology-concentrated geographies, innovation and knowledge creation hubs, and policy and planning implications of innovative technologies. This makes the special issue particularly of interest to those engaged in smart city and innovative urban technology research, practice, and policy making.

The first paper of the issue, by Nicos Komninos, Christina Kakderi, Anastasia Panori, and Panagiotis Tsarchopoulos ("Smart City Planning from an Evolutionary Perspective"), extends the evolutionary thinking and emerging dynamics of cities to smart city planning. The paper uses Thessaloniki, Greece as a case study to reveal a smart city strategy that has the potential to enhance the economic, environmental, and social sustainability of a city. The study reveals the complex dimension of smart city planning as a synthesis of technologies, user engagement, and windows of opportunity, which are fuzzy at the start of the planning process.

In the second article, Tan Yigitcanlar and Md Kamruzzaman ("Smart Cities and Mobility: Does the Smartness of Australian Cities Lead to Sustainable Commuting Patterns?") concentrate on smart mobility. The paper investigates smart mobility from the angle of sustainable commuting practices in the context of smart cities. The study reveals that overcoming the need for car-based travel for fragmented work activities while increasing smartness through the provisioning of broadband access should be a key item on the smart city agenda.

The security vulnerabilities of smart cities are examined in the third paper of the special issue by Rob Kitchin and Martin Dodge ("The (In)Security of Smart Cities: Vulnerabilities, Risks, Mitigation, and Prevention"). The study adopts a normative approach to explore existing mitigation strategies, suggesting a wider set of systemic interventions. It reveals how this approach might be enacted and enforced through market-led and regulatory measures, and then examines a more radical preventative approach to security.

The issue continues with the fourth paper by Juho Kiuru and Tommi Inkinen ("E-Capital and Economic Growth in European Metropolitan Areas: Applying Social Media Messaging in Technology-Based Urban Analysis"). The research explores European metropolitan areas in relation to their Twitter messaging activity. The study applies a quantitative approach combining social media activity with regional economic variables. The paper reveals the most social-media-intensive locations in Europe in relation to innovation and start-up messaging.

In the fifth paper, by Luca Mora, Mark Deakin, Alasdair Reid, and Margarita Angelidou ("How to Overcome the Dichotomous Nature of Smart City Research: Proposed Methodology and Results of a Pilot Study"), introduce a research methodology for conducting the multiple-case study analyses, and tests the practical feasibility, effectiveness, and logistics of such a methodology by examining the smart city initiative of Vienna. The findings reveal how the application of the proposed methodology can help smart city researchers codify the knowledge produced from multiple smart city experiences, using a common protocol.

The sixth paper, by Becky Loo and Winnie Tang ("'Mapping' Smart Cities"), focuses on a systematic review of the development and functionality of digital maps. The study uses six domains depicting smart cities and reflects the digital map development against them. The study discloses that progress in digital mapping has supported and enabled smart city initiatives. The findings of the study stress the importance of smart mapping as an essential step in promoting smart cities and highlight mapping's usefulness in supporting smart cities initiatives.

The seventh and final paper, by Tan Yigitcanlar, Marcus Foth, and Md. Kamruzzaman ("Towards Post-Anthropocentric Cities: Reconceptualizing Smart Cities to Evade Urban Ecocide") acts as a coda to the special issue. This study provides a retrospective view of the origins of the smart city concept, presents the most recent perspectives on new interpretations of the smart city notion, and provides a commentary on the potential directions for a better reconceptualization of smart cities to evade a most likely urban ecocide.

We cordially thank the contributors of the special issue for their high-quality research papers and the editorial staff of *Journal of Urban Technology* for their support in preparing the special issue. We hope that this issue brings forth new ideas and stimulates future research interests in the fields of urban planning, civil engineering, urban development, economic geography, urban technology, and other related disciplines. We also express our warm appreciation to the expert scholars, who participated in the double-blind peer-review process and provided their constructive comments and insightful suggestions for improvement. Finally, we thank the Editor-in-Chief, Richard Hanley for giving us the opportunity to edit this special issue.

Tommi Inkinen

Tan Yigitcanlar

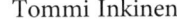

Mark Wilson

Smart City Planning from an Evolutionary Perspective

N. Komninos ⓘ, C. Kakderi ⓘ, A. Panori ⓘ, and P. Tsarchopoulos

ABSTRACT

In the theory of urban development, the evolutionary perspective is becoming dominant. Cities are understood as complex systems shaped by bottom-up processes with outcomes that are hard to foresee and plan for. This perspective is strengthened by the current turn towards smart cities and the intensive use of digital technologies to optimize urban ecosystems. This paper extends the evolutionary thinking and emerging dynamics of cities to smart city planning. It is based on recent efforts for a smart city strategy in Thessaloniki that enhances the economic, environmental, and social sustainability of the city. Taking advantage of opportunities offered by the IBM Smarter Cities Challenge, the Rockefeller 100 Resilient Cities, the World Bank, and the EU Horizon 2020 Program, Thessaloniki shaped a strategy for an inclusive economy, resilient infrastructure, participatory governance, and open data. This process, however, does not have the usual features of planning. It reveals the complex dimension of smart city planning as a synthesis of technologies, user engagement, and windows of opportunity, which are fuzzy at the start of the planning process. The evolutionary features of cities, which until now were ascribed to the working of markets, are now shaping the institutional aspects of planning for smart cities.

Smart Cities from an Evolutionary Perspective: Frame of Reference

Masdar City is a landmark in twenty-first-century urban development as it is the first zero carbon city, opening up an era of technology-led sustainability and green growth. But, is Masdar a city? According to *The Guardian* (Goldenberg, 2016) only 300 people so far live on the site and all are students at the Institute of Science and Technology. In fact, Masdar is actually a group of buildings, a large physical complex; more an engineering construct than a city. It will become a city in the future, when people and human activities, culture, institutions, and behaviors give purpose and use to infrastructures and buildings. Masdar will evolve into a city, as all cities do; they evolve and become cities rather than being constructed as cities from scratch. This idea of "cities becoming cities" rather than "cities planned as cities" is a core premise of evolutionary thinking about urban development. Cities are extremely complex and chaotic systems; many forces work simultaneously in their making and even small variations in the outcome interact and produce huge changes in results. Economic and political forces create numerous constraints on cities, yet there is room for genuine development that is not bound by deterministic conditions.

Evolutionary thinking holds a preeminent position in urban and regional development theory. Cities and regions offer resources that are actualized by selective mechanisms that drive change and growth. Lambooy (2002) argues that urban regions offer effective contexts for development through an evolutionary process where cognitive, innovative, and organizational competencies are influenced by a selection environment composed of institutions, markets, and spatial structure. This environment drives the choice between alternative planning ideas and designs for new investments in city services and infrastructures. Here there is an analogy to the way Nelson and Winter (1977) have described innovation as a purposive, but inherently stochastic activity, which is guided by an external selection environment that determines how different technologies are selected and change over time. The innovation selection environment is shaped by market and non-market forces, consumer preferences, investment, and imitation processes, as well as political and regulatory control over firms. Simmie and Martin (2010) widen this understanding of how innovation in cities is produced, connecting the development of cities and regions to four conceptual frameworks that offer an evolutionary account of resilience and adaptation: (1) generalized Darwinism which places emphasis on variety, novelty, and selection; (2) path dependence theory that underlines historical continuity "lock-in" and new path creation; (3) complexity theory with its emphasis on self-organization, bifurcations, and adaptive growth; and (4) panarchy that links resilience and "adaptive cycles." Boschma (2004) points out the uniqueness of urban and regional growth paths from an evolutionary perspective, since the competitiveness of a region depends on intangible, non-tradable assets resting on a knowledge base embedded in the region's specific institutional setting. Transferring growth models from one region to another is questionable as there is no "optimal" development model, and new successful trajectories and developmental paths emerge spontaneously and unexpectedly in space. Bettencourt et al. (2010) argue that agglomeration non-linearities connect most urban socioeconomic indicators with population size, making larger cities centers of innovation, wealth, and crime. They find that local urban dynamics display long-term memory, so cities under- or out-perform their size expectation and maintain such advantage for decades.

All the above statements are meaningful for smart city planning: a process that highlights the uniqueness of each city trajectory, is based on rapidly changing digital technologies, and is ready to value opportunities offered over time rather than copycat planning, locked-in optimal models and one-size-fits-all solutions. The case study we discuss in this paper presents a decision-making environment in a state of constant change, which is discontinuous and non-linear, but offers unexpected windows of opportunity; a complexity that has few commonalities with spatial planning as an ordered process that guides actions from an existing situation to an envisaged future (See also De Roo and Silva, 2016). The scientific ambition of the paper is to reveal the evolutionary dimension of smart city (or intelligent city)[1] planning, due to rapidly changing digital technologies and opportunities that in many cases do not exist at the start of the planning process, which justify the need to replace rigid and well-defined city plans with roadmaps that enable them to integrate evolving technologies and initiatives.

Thus, in this paper, we expand the evolutionary perspective of urban growth to smart city planning. We argue that due to the complexity of smart city development processes and the multi-disciplinary character of smart city technologies, smart city planning is shaped by evolutionary processes too. Evolutionary processes are characterized and affected by essential diversifications in the capacity of societies to generate technical innovations that are suitable to their needs (Rosenberg, 1982). These differences also relate to higher complex systems of policy design that form pools of opportunities for funding and research. Cities and urban planning processes are affected by these dynamic environments, when trying to efficiently exploit existing opportunities for policy formation, in order to achieve a leading position within the global context, to attract more funds and inward investment. It is important to understand that urban and regional developmental evolutionary paths depend on the nature of selection environments, such as public funding, administrative rules, policy frameworks, and others. In this case, the selection process is shaped by political, economic, and cultural factors and the competencies of carrying actors and institutions (Lambooy, 2002). Urban contexts influence the ways in which local governments can create and shape opportunities for innovation.

Planning for smart cities—or the use of digital technology to innovate and improve urban ecosystems—has become a major strand of contemporary urban planning literature. Since the beginning of 2017, publications on smart cities have accounted for close to 50 percent of all publications related to urban planning (Google Scholar data). Yet, major aspects of this new planning model are not well understood, especially the interaction between and integration of long-term, top-down plans and short-term, bottom-up initiatives.

The planning objectives and the type of smart city projects that cities implement are also highly diverse (Yigitcanlar, 2016). Take for instance, three well-known cases of smart city strategy: Singapore Intelligent Nation, Amsterdam Smart City, and Smart Santander. A sector-focused approach in Singapore is implemented using web-based platforms in the domains of digital media, financial services, manufacturing, logistics, and others, compared to projects focusing on sustainability, energy savings, CO_2 reduction, and user participation in Amsterdam, and the deployment of numerous sensors and Internet of Things infrastructure in Santander over which technology providers are asked to develop applications and e-services. These cases illustrate very diverging approaches both in terms of planning priorities and the understanding of how smart cities work.

To our mind, smart city planning defines a distinct phase in the evolution of urban planning, a new planning paradigm that differs substantially from the Twentieth Century and mainly the post-WWII schools of planning (Hall, 1988). This perspective nurtured the discussion about a new science of cities (Batty, 2013; Bettencourt and West, 2010) with cities seen as entities that enable communication and networking, and therefore producing externalities for wealth and the saving of infrastructure, regardless of the economic and geographical context. However, the critical factors that clearly differentiate smart city planning from previous planning perspectives are the knowledge base and the mode of operation. The City Beautiful movement and the plans of Haussmann in Paris, Burnham in Chicago, Lutyens and Baker in New Delhi, Griffin in Canberra, and Hébrard in Thessaloniki were based on knowledge supplied by engineering sciences, architecture, and landscape design. Later, throughout most of the Twentieth Century, the modernist movement for the rebuilding of urban centers and/or suburban sprawl was based on understanding the role of the state in urbanization, regulations and policy incentives for urban development and building, control of land uses, creation of large-scale infrastructure for mobility, social housing, and welfare economics; in sum, a knowledge base provided by social sciences, theories of location, land and traffic management, and strategic planning. Currently, the making of digital, smart, and intelligent cities, uses different materials, such as broadband communication networks, sensors, big datasets, software applications, and e-services. Their knowledge base is offered by programming languages, algorithms, mining large datasets, analytics, software design and development, and user engagement and co-design. This historical expansion of city planning's knowledge base has been cumulative and interdisciplinary with each subsequent field of knowledge adding new elements to the previous one, but also retaining most of the previous theoretical construction.

Planning for smart cities starts with the creation of the urban digital space, an agglomeration of digital hardware and software, datasets from the public administration, sensors and smart meters, social media, and new e-services in every domain of the city. This new layer of digital space and technologies has the capacity to change and optimize all aspects of cities: the economy, life, utilities, and governance. We have called this process "innovation circuit 1" (IC1) which creates the digital space of cities. The overall smart urban system is made of heterogeneous and uncoordinated initiatives by the public administration, global social media companies, national telecom companies, IT developers, e-service providers, and users; each actor adding some digital component to a common pool of resources, and each one offering new modes of user engagement, participation, and empowerment. In parallel to the formation of the urban digital space, two other processes of innovation emerge: more informed decision-making and governance of public and private investments that drive the change of cities ("innovation circuit 2" [IC2]); and more efficient citizen behavior based on urban awareness that guides the use of urban space and infrastructure through intelligent systems, GPS, and sensor-based solutions ("innovation circuit 3" [IC3]) (Komninos, 2014, 2016b). These three circuits, taken together, define smart city planning and describe the operation of smart or intelligent cities as complex cyber-physical systems of innovation. Innovation circuits 2 and 3 are based on and become possible thanks to the digital space of cities. Innovation circuits IC1, IC2, and IC3 work in tandem; there is no evolution among them. They occur simultaneously; the moment IC1 is introduced, depending on its functionality, it enables better decision-making and / or optimized user behavior. When IC1 relies on web 2.0 technologies, collaboration platforms or crowdsourcing solutions, decision-making becomes participatory with the engagement of users. They constitute forms of citizen empowerment and data awareness, either by the city producers or the city users.

Understanding the planning and making of smart cities through the juxtaposition of digital elements, which are heterogeneous, uncoordinated and usually not integrated, and through novel producer and user behavior, which is also fragmented and diverse, is far from the usual concept of urban planning we have been used to. Thus, smart city planning, as control and guidance of the entire interaction between innovation circuits 1, 2, and 3, is "planning without a plan," and the making of cities through evolution rather than through detailed design and rigid plans. It is planning under uncertainty, chaotic interaction of concurrent actions by many organizations, each one having its own rationality and plan. Or, planning by the same organization guided and shaped by opportunities that appear over time, with the overall outcome being unpredictable and uncontrolled at the beginning. Smart city technologies and their impact on innovation systems are the main causes for this trajectory.

The ambition of the present paper is to bring up and reveal the uncertain aspect of smart city planning, as an agglomeration of initiatives and actions, and windows of opportunity, which are uncoordinated and unpredictable. To our mind, this feature is not a side effect of some ill-designed planning process, but a structural result of the core drivers of smart cities, namely, the modalities of digital/smart space, the availability of large datasets, extended citizen empowerment in city decision-making and design, and the creation of cyber-physical systems of innovation (Komninos, 2016a).

Following this introduction on the topic and frame of reference, the rest of the paper consists of four sections. The next section refers to the evolution of smart city technologies, outlining the main stages and milestones. Technologies from broadband to sensors, datasets and applications, and their interdependencies constitute a critical dimension of smart city complexity. Then, we focus on a case study: smart city planning in the city of Thessaloniki over the last five years, guided by local and global initiatives, such as the Rockefeller 100 Resilient Cities, the IBM Smart Cities Challenge, Horizon 2020 research, and others, which illustrate the evolutionary character of smart city-making. The last two sections discuss findings from the technology landscape and the case study as instances of an evolutionary model for smart city planning, its core features, and their implications for the future of cities.

Evolution of Digital Technologies: The Foundation of Smart City Complexity

Understanding the operation of intelligent cities *through* the three innovation circuits (IC1, IC2, and IC3) mentioned above, places the origin of smart cities in the digital space that sustains citizen innovative behavior, and more informed investment and governance practices. The digital space of cities is created by a large variety of elements, such as broadband networks, sensor networks, urban operating systems, web spaces, datasets, and urban informatics. It can be described by a series of layers or rings, each one having specific characteristics and functionality: (a) broadband networks, wired and wireless infrastructure, and communication protocols enabling communication and the connectivity of various devices embedded into the urban space; (b) data creation and collection technologies, such as sensors, smartphones, actuators, (c) databases, algorithms, and programming languages, which allow for dataset creation and processing, data visualization, and analytics; (d) web and smartphone technologies enabling the creation of applications with functionality targeted to different domains of the city; at least 20 different domains of cities can be identified as potential fields of applications related to the economy, city infrastructure and utilities, quality of citizen life, and city governance (see the ICOS software repository at icos.urenio.org); and (e) e-services addressed to citizens and organizations, based on applications adopted by the market and offered on a regular basis as a service via viable business models. In a condensed and articulated form, all these elements can be found in the so-called "urban operating systems" which integrate network infrastructure, sensors, devices, software applications, and people across different domains and urban systems (Marvin and Luque-Ayala, 2017; Living Plant, 2016).

This complex digital edifice of cities has been created gradually *through* the accumulation of technologies, smart systems and solutions, and to a large degree it follows and depends on the progress of the Internet and the world-wide-web. We can identify three successive phases or waves of development, each one linked to specific technologies and features of the corresponding digital space.

The first wave of smart city solutions concerned the representation of the city, in early forms via portal-type webpages, panoramic and 3D representations of cities, and later via augmented reality technologies, and urban tagging. Digital cities are connected communities that combine "broadband communications infrastructure; a flexible, service-oriented computing infrastructure based on open industry standards; and, innovative services to meet the needs of governments and their employees, citizens and businesses" (Yovanof and Hazapis 2009: 446). Digital cities tried to link the physical and digital space by offering a metaphor of the city; an understanding of the city through its virtual representation. Such digital cities were described as "mirror-city metaphors" or "virtual cities," as their logic was to offer "a comprehensive, web-based representation, or reproduction, of several aspects or functions of a specific real city, open to non-experts" (Couclelis, 2004: 5). Differences in representation models resulted in differences in functionalities, which ranged from simple, informative webpages, to communication spaces with forums and chatrooms, and finally to interactive spaces with virtual agents. The spatial intelligence of cities related to digital solutions of this type was based on the advantages of representation and visualization. The expression "one picture is worth a thousand words" reflects this idea that complex environments can be described and understood better by a virtual representation or metaphor. In the field of theory, the digital city literature benefited from the work of Ishida and Isbister (2000), Hiramatsu and Ishida (2001), and Van den Besselaar and Koizumi (2005). The solutions mentioned above were content-intensive and required fast Internet connections; thereby they encouraged the adaptation of broadband access by the city's population. Telecommunication companies started creating new backbone networks for data exchange using fiber optics and xDSL technologies, while the city authorities began to build local wireless networks.

Advances in broadband connectivity (wired and wireless) combined with the arrival of the Web 2.0 concept (O'Reilly, 2007) catalyzed the evolution of smart city solutions. In the second wave of smart city solutions, the focus shifted from the representation of the city to solutions that enabled citizen participation and engagement in smart city creation. The rise of the social web led to the creation of digital spaces that harnessed citizens' collective intelligence to organize the development of technologies, skills, and learning, and to engage citizens to become involved in creative community participation (Deakin and Allwinkle 2007). Co-creation and crowdsourcing were the most common forms of collaboration in the second wave of smart city solutions. City intelligence came onto the scene with the understanding that digital spaces improve urban ecosystems by processing information, sustaining learning, and innovation produced by user engagement and networks of collaboration. It emerges from a combination of the creative capabilities of the population, knowledge-sharing institutions, and digital applications organizing collective intelligence. Within cyber-physical urban agglomerations, forms of distributed intelligence connect (a) the inventiveness, creativity, and human intelligence of the city's population, (b) the collective intelligence of the city's institutions and social capital for innovation, and (c) the artificial intelligence of public and city-wide smart infrastructure, virtual environments, and intelligent agents (Komninos, 2008). From this perspective, in the second wave of smart city solutions, the spatial intelligence of cities was built on collective intelligence and social capital for collaboration, combined with a people-driven innovation introducing principles of openness, realism, and empowerment of users in the development of new solutions (Bergvall-Kåreborn and Ståhlbröst 2009). The ever-increasing participation of citizens in smart city solutions has been facilitated by the adoption of cloud computing, which disengages city authorities from resource constraints, whether they are technical, managerial, or financial. Cloud computing has a higher impact and greater effect at the city level, as it enables city authorities to create a highly efficient, scalable, and elastic computing environment for smart city service provisioning (Kakderi et al. 2016).

In a third and more recent turn, the interest in smart cities is sustained by two new concerns: on one hand there is the rise of new Internet technologies promoting real-world user interfaces via mobile phones, smart devices, sensors, RFIDs, the semantic web and the Internet of Things, and on the other there is the concern for sustainability and how smart cities can support a more inclusive, diverse, and sustainable urban environment, green cities with less energy consumption and lower CO_2 emissions (Caragliu et al., 2011). Currently, the smart city literature focuses on the latest advancements in mobile and pervasive computing, wireless networks, middleware, and agent technologies as they become embedded into the physical spaces of cities and are fed with data round the clock. Smart city applications—with the help of instrumentation and interconnection of mobile devices and sensors that collect and analyze real-world data—improve the ability to forecast and manage urban flows and push city intelligence forward (Chen-Ritzo et al., 2009). Within this technology stack, spatial intelligence moves out of applications and enters into the domain of data: the meaning of data becomes part of data, data are provided just-in-time, and real-time data enable real-time response. Artificial Intelligence (AI) is a perfect fit for this new situation of smart city systems. As smart cities gather a significant amount of data, AI can provide tools and techniques to analyze them and get insights hidden into data. It can detect emergent patterns. It enables multiple systems to be optimized together, and provides entirely new capabilities that traditional analytics tools cannot. Moreover, through deep learning and natural language processing techniques, it enables new modes of human–machine interactions, making access to smart city solutions easier and in real time.

These changes were extremely rapid and the outcomes take us closer to Ambient Intelligence Environments. From a technology perspective, Ambient Intelligence combines broadband and sensor networks, processing power, reasoning mechanisms, applications and e-services embedded into the surrounding environment. It represents a vision for the future where intelligent or smart systems interact with citizens in an adaptive way that sustain humans living and working within urban environments (Streitz, 2017).

Smart City Planning in Thessaloniki: Taking Advantage of Windows of Opportunity

But technology is not sufficient on its own to explain the evolutionary making of smart cities, which is also guided by user engagement, flexible governance, business models, investment opportunities, and other initiatives for city improvement. The case study on the city of Thessaloniki that we discuss in this section shows how technologies and planning complement each other in valorizing opportunities for smart city development, which appeared gradually, without coordination, both locally and globally.

Thessaloniki is the second largest city in Greece with a population of over 1,100,000 (in the metropolitan area). The city has made significant efforts to implement a number of activities that contribute to its journey towards becoming a smart city. In terms of strategic design, the first comprehensive plan for creating a smart city in Thessaloniki was prepared by URENIO Research, a lab of the Aristotle University, in cooperation with the Regional Government of Central Macedonia in 2009. "Intelligent Thessaloniki" was a strategy to strengthen the city's innovation ecosystems through the deployment of open public broadband networks and the development of web applications and smart environments. The strategy focused on selected city districts and production ecosystems (CBD, port area, university campus, innovation zone) in which broadband networks and a wide range of digital applications and e-services, tailored to each district's characteristics, were proposed to improve innovation capabilities and entrepreneurship (Komninos and Tsarchopoulos, 2013).

In the years that followed, the Intelligent Thessaloniki strategy was not implemented. A change of government and an overwhelming financial crisis were the main reasons for this project being abandoned. But, digital Thessaloniki continued to emerge bottom-up as an agglomeration of commercial and community broadband networks and web-based services for government, education, business, mobility, quality of life and other activities of the city. These were fragmented and independent efforts made not only by large telecommunication companies, Internet service providers, and ICT companies, but by civic communities, small IT companies, and individual developers.

Since 2013, the Municipality of Thessaloniki has taken the lead as the implementing agency for efforts to create a smart and resilient city. The Municipality agreed to become an active partner of Aristotle University of Thessaloniki and managed to garner strong support from all stakeholders in the city. The first outcome of that collaboration was the STORM CLOUDS project (Surfing Towards the Opportunity of Real Migration to cloud-based public services); a research project that was partly funded by the European Commission in the context of the Competitiveness and Innovation Framework program (CIP PSP).[2] The project, which started in February 2014 and ended in March 2017, aimed to accelerate the pace at which public authorities move to cloud computing. Thessaloniki was among the four pilot cities, with an emphasis on the smart economy. The project introduced the concepts of smart cities and cloud computing as a disruptive model for the uptake of smart city services to the Municipality's administration and personnel. Subsequently, the Municipality began to take similar initiatives, with the most noteworthy being participation in the European Commission's "Innovation Partnership on Smart Cities and Communities"[3] and the Smart Cities MoU with the largest Greek cities (Athens and Heraklion).

The collaboration between the Municipality and Aristotle University had already begun earlier with the organization of the first smart cities app contest in Greece, named "Apps for Thessaloniki,"[4] jointly with the Greek Chapter of the Open Knowledge Foundation (OKF Greece). The competition ran for five months (November 2013–March 2014) and aimed to stimulate the local ICT ecosystem to create new smart city solutions. Thirteen applications were developed, covering a wide range of city domains and activities. The following year (November 29–30, 2014) another smart city contest took the form of a "hackathon" in which 10 teams of developers participated.[5] The concept has evolved into a thematic competition targeting specific city domains (i.e., tourism, energy, environment, etc.). The Apps for Thessaloniki--Tourism edition (November 2015–January 2016) produced 12 web and mobile applications.

Open Data

The involvement of the Municipality of Thessaloniki in apps contests and hackathons, as well as the collaboration with OKF, resulted in the creation of a movement for open data and open government within the Municipality. The newly established Department for e-Government released the first open datasets in November 2013. Over the years that followed, the open data movement has strengthened. Regarding open government, a dedicated portal was released in 2015, which allows citizens and businesses to access most of the Municipality's services through web or mobile applications.[6] "Improve my City" (Tsampoulatidis et al., 2013), the portal's flagship application that allows citizens to submit and comment on non-emergency problems related to the urban environment is used by thousands of people, who in this way contribute to improving the city while also engaging with the Municipality. In May 2017, this application received an Award from the Council of Europe at the "European Label of Governance Excellence" opening event for digital services provided by the Municipality of Thessaloniki.

The commitment to open data has paid off as the City of Thessaloniki was selected through a competitive process as one of 16 cities to be awarded a Smarter Cities Challenge grant in 2015 to 2016 by IBM.[7] With IBM's support, the city seeks to integrate diverse open data sources across the fields of governance, mobility, education, environment, and economy. In February 2017, the IBM Smarter Cities Challenge team published a report containing recommendations and a roadmap that will help the city to achieve this goal.

Establishing Thessaloniki as a leader in open data is a priority for the City's Mayor Yiannis Boutaris. To that end, the Municipality entered into a strategic partnership with OKF. Initially, the Municipality released a few datasets containing spatial data to support the smart city app contests and hackathons. Currently, the portal contains 74 datasets in eight categories: Urban Planning (36), Public Administration (13), Environment (13), Tourism (6), Education (5), Culture (4), Public Security (1) and Economy (1).[8] Moreover, the city publishes open data regarding budget spending. Citizens can monitor implementation of the city's budget in real time and use visualization tools to have a better understanding of the budget data.

Thessaloniki participated in the IBM Smarter Cities Challenge with a proposal in this field. The challenge for the city was to "develop a strategy and tactics that will help the City utilize open data to encourage further transparency, benchmarking, key performance indicators (KPIs) and data-sharing between public departments, businesses, universities, non-governmental organizations (NGOs) and citizens" (IBM, 2017). Following the award of the grant, during a three-week period in November 2016, a team of six IBM experts worked in Thessaloniki to deliver recommendations on open data infrastructure and organization. The team conducted more than 40 interviews with various stakeholders (public office holders, City employees, university faculty members, local entrepreneurs, and leaders of NGOs) across the City's ecosystem. In February 2017 IBM's team presented strategic recommendations to advance open data adoption under the following five themes:

(1) Reorganize IT-related departments to enable open data policies and practices.
(2) Establish an open data strategy and consistent understanding across City departments and stakeholders.
(3) Foster an environment that supports collaboration.
(4) Establish a publishing process and maturity model that put open data into practice.
(5) Address resource constraints through investments, strategic partnerships, and change management.

Moreover, IBM proposed the development of an open data dashboard combining data from different stakeholders and providing citizens, public sector employees, and companies with real-time information, time-series data, and interactive maps about all aspects of city life. The city dashboard will enable users to gain detailed, up-to-date intelligence about the city for daily decision-making and evidence-informed analysis.

Currently the City of Thessaloniki is the leader among Greek cities regarding open data and administration transparency. It ranks in first place in the Greek Cities Open Data Census run by the Open Knowledge Foundation.[9] With the creation of the city data dashboard, the city will put open data to work for its residents.

Collaborative Economy

Over the last decade Thessaloniki has been hit by the economic crisis while its image has been affected significantly by attempts at corruption in various departments of the Municipality. A change in local government was accompanied by efforts to rebuild trust but also to improve the economic environment for business and investments in the city. The municipality participated in various projects funded by the Horizon 2020 Program for the development of digital services related to entrepreneurship and the promotion of tourism.

By adopting a user-driven methodology through meetings and workshops with stakeholders and municipal services, the city launched two popular applications: the "Virtual City Market" and the "City Branding." The Virtual City Market is an application that, on the one hand, enables every commercial enterprise located in the city to create its own virtual shop and, on the other, enables customers to access a variety of retailers using a shared site. In its simplest form, the service provides a list of existing shops located in the city (and their location on a map) as well as what they offer. The Virtual City Market enhances collaboration schemes between retailers, offering the opportunity to create open malls and organize the shops per street or district.

City branding is an application that promotes the identity of a city to different target groups using virtual tours and presentation of points of interest, while being connected to the local economy and entrepreneurship. The application allows a city to focus on different target groups that are associated with various aspects of the city's identity (history, culture, economic environment, etc.) by supporting the differentiation of commons according to target groups of visitors. Both the above applications have been funded by the STORM CLOUDS project.[10]

Moreover, in collaboration with the city's universities and business associations, in 2016 the Municipality launched the "OK!Thess" initiative. OK!Thess is an innovation ecosystem for startups that offers a temporary working space to newly established enterprises together with activities such as training, consultation, organization of networking activities, the search for funding, and the promotion of product and services.

Sustainable and Resilient Infrastructure

Another major international distinction for the city was its participation in the Rockefeller Foundation 100 Resilient Cities initiative (100RC)[11] dedicated to helping cities around the world become more resilient in the face of physical, social, and economic challenges. Thessaloniki was selected in 2014 as part of the second cohort of cities to join the 100RC network.

Over a period of two years, more than 2,000 people and 40 organizations from across the city were engaged in the design of the strategy, participating in workshops and filling in questionnaires expressing their views on Thessaloniki's resilience. In addition, to maximize its added value, there was research on more than 1,000 actions undertaken by the Municipality and a comparative study of more than 700 completed or institutionalized plans in the Municipalities that comprise the Thessaloniki Metropolitan area. In line with the main problems faced by the city, the focus was on issues related to the local economy and mobility.

In 2016 a Deputy Mayor of Urban Resilience and Development Planning and Chief Resilience Officer was appointed. The Resilience Strategy for Thessaloniki, published in March 2017, is the first city-wide collaborative strategy and at the same time is a roadmap in the city's effort to guarantee the well-being of its citizens, to nurture its human talent, and to strengthen the urban economy while respecting its natural resources. The strategy is built on four main goals, 30 objectives, and more than 100 actions (Thessaloniki, 2017). The goals of the strategy are: (a) shaping a thriving and sustainable city with mobility and city systems that serve its people; (b) co-creating an inclusive city that invests in its human talent; (c) building a dynamic urban economy and responsive city through effective and networked governance; and (d) re-discovering the city's relationship with the sea—integration with Thermaikos Bay.

At present, the Municipality is in the process of implementing the strategy leading a new round of consultation with the city's stakeholders to get support and encourage engagement in different aspects of the implementation process (leverage funding, selection of KPIs, collection of open data, etc.). As Thessaloniki has a huge reserve of youth (with more than 100,000 students in the three higher education institutions located in the metropolitan area), a dialogue has also opened with young people living in the city[12] and with the academic community through various initiatives, workshops, and events.

In pursuit of this goal, the Municipality signed a memorandum of understanding (MoU) with the World Bank in May 2017. The MoU describes the offer of technical assistance by the World Bank in the strategic design and development of policies and programs related to issues of mobility and transport, resilience and crisis management, economic and urban development, and investments attraction, etc.

Participatory Governance

Participatory governance is also a core concern of the current administration, fostering the efforts for sustainability and a place-based governance model. Given the efforts that have been made towards a resilient Thessaloniki, the development of a participatory governance infrastructure emerged as a natural outcome. In this context, the *ImproveMyCity* and *CloudFunding* applications were developed to strengthen the effectiveness of the local governance framework, resulting in *more informed decision-making processes* (in innovation circuit 2), aiming to drive changes within cities towards a sustainable urban space and towards local social inclusion.

ImproveMyCity is an application driven by the intriguing concept that every citizen can act as a living sensor in the city. The participatory governance concept is the cornerstone of this application; a concept which has been developed under the CIP project PEOPLE.[13] Overall, the platform provides a user-friendly interface where citizens can directly report non-emergency issues about their city, indicating the exact location on the map, as well as the nature of the problem. Users can add photos and comments. In this way, citizens can become local actors themselves, suggesting solutions about how to improve the environment of their neighborhood (Tsampoulatidis et al., 2013).

In terms of administration, this application helps local authorities organize the reported cases for further action and resolution. There is live information regarding the time frame for resolving the reported issue, while the person who originally submitted the request is directly informed about the outcome. Another additional feature of the application offers the administrator the opportunity to visualize data and identify specific areas with a high share of dissatisfied citizens or under-performing administrative departments. Overall, the *ImproveMyCity* application works as a means for strengthening governance procedures via citizen participation and thus, urban sustainability, through the improvement of public space. Citizens are an active part of this process, defining the main issues that need to be solved at a local level, thereby reinforcing the participatory governance of the city.

CloudFunding is different. It is a platform that supports civic crowdfunding and has been developed under the STORM CLOUDS project. Through this application, local authorities can support communities in collecting money for social and charitable purposes. Supported projects refer to urban sustainability and three types of initiatives related to: (a) improvement of the city's physical environment (i.e., the creation of parks and playgrounds, restoration of monuments, expansion of cycle lanes, etc.), by combining private and public funding; (b) social entrepreneurship (i.e., creation of non-profit enterprises to promote objectives that improve urban life or strengthen the city's social capital), in which case local authorities will act as a mediator of the initial effort; and (c) knowledge-intensive and technology-based youth entrepreneurship. In all cases, the user has to define a minimum and an optimum target for the desired co-financing, as well as the period over which this project will run. Each project has to make clear what the benefits to the local community will be, and this must be clearly shown on the platform.

In both these cases, participatory governance is expressed through the ability of citizens to engage and provide information about specific issues related to their local communities. Citizens define and become aware of local actions, a fact that enables them to significantly contribute to the overall sustainability of the city. Projects and local issues are classified based on public opinion, and thus, their implementation is driven by the overall social benefit for the local community.

Discussion: Smart City Planning without a Plan

The case study of Thessaloniki indicates the multidimensional character of actions that are deployed to transform a city into a smart and sustainable place. Smart city sustainability refers to a set of dimensions, including socioeconomic, environmental, and governmental dimensions, which can be enhanced through the use of smart city applications, networks, and integration of digital, social, and institutional elements. Given the fact that the Municipality of Thessaloniki has taken the lead as the main agent for promoting and implementing efforts to create a smart and resilient city, its active collaboration with Aristotle University of Thessaloniki has established an effective channel to accumulate support from a large number of stakeholders in the city.

There have been many parallel efforts and initiatives to promote this vision, including ICT solutions fostered by civic communities and individual developers. Starting from EU co-funded projects, such as the PEOPLE and STORM CLOUDS projects, and moving on to the organization of hackathons and OK!Thess, and collaboration with organizations such as IBM and the Rockefeller Foundation, all attempts have tried to stimulate the local ICT ecosystem in order to create a new set of smart city services and synergies. These work as accumulative forces towards a collaborative resilience-building process, thereby fostering urban sustainability.

At the same time, openness and inclusion have been strengthened through the development of applications related to local economic activities. City Branding and Virtual City Market are both perceived as ways to enhance the links between citizens and local opportunities, in terms of activities and of market infrastructure. This leads to the development of a sustainable urban economic environment, where people are well informed about existing opportunities.

ImproveMyCity and *CloudFunding* are two cases highlighting the efforts of the Municipality of Thessaloniki to promote and encompass participatory governance throughout the decision-making processes. Citizen participation in defining requests and priorities, as well as assessing possible benefits from funding place-based projects, provides a valuable source of information. This enriches urban sustainability, as it offers local authorities the opportunity to strengthen their effectiveness, by incorporating public opinion in their planning processes.

Open data initiatives are also at the core of the city's strategic efforts, making it the leader among Greek cities in this regard. The open data portal of the Municipality of Thessaloniki is considered to have been a focal point throughout this overall transformative process, reinforcing data openness and transparency in a wide set of categories. The IBM Smarter Cities Challenge has also been a milestone for Thessaloniki during this process, leading to an open data city dashboard, combining input data from different stakeholders.

The case study on Thessaloniki presented here clearly shows that the strategy and actions guided by the vision for an open, global, smart, and resilient city, have been largely shaped by a series of opportunities that appeared gradually over the last few years, both at global and local levels: the Rockefeller Initiative for 100 Resilient Cities, the IBM Smarter Cities Challenge, the collaboration with the World Bank, the collaboration with Aristotle University in Horizon 2020 projects, the need for digital strategies for getting access to European Structural and Investment Funds (ESIF) funding, as well as collaboration with the OKF in hackathons and software competitions. These initiatives have defined the framework for guidance, know-how, funding and citizen engagement, and have shaped a smart city planning approach which was neither top-down nor defined in advance. Actually, there is a strategy and an action plan for a smart city (Thessaloniki, 2017; IBM, 2017; Municipality of Thessaloniki, 2017), but they were formed gradually, in an evolutionary way, through the convergence of independent initiatives and the specific frameworks and goals of those initiatives.

Conclusions: Towards an Evolutionary Perspective of Smart City Planning

This smart city strategy and action plan formation process challenges not only the concept of top-down planning, but also the capacity for smart city plans being formulated exclusively by state authorities. Smart city planning as a complex process was discussed by Leydesdorff and Deakin (2011) and Deakin (2015). The authors link smart city planning to the rise of triple helix governance and attribute its neo-evolutionary character to three functions that shape the selection environments of the smart city knowledge economy: organized knowledge production, economics of wealth creation, and reflexive control. Reflexivity is not a given, but socially constructed by evolving communication systems and cultural settings.

No doubt, the triple helix is a driver of complexity. All the more so is quadruple helix governance with the wide participation of users and multi-actor decision-making. The evolution of technologies and the case study discussed earlier reveal that strong drivers of complexity are also the innovation push created by initiatives launched by global organizations, bottom-up innovation introducing applications and e-services, and the changing urban behavior of users due to real-time information and participation through social media. Cities take advantage of initiatives, partnerships, and policy frameworks at regional and national levels also, which evolve over time, appear as windows of opportunity, and disappear after a while to give way to other opportunities. At a regional level, for instance, the search for investment opportunities is expressed by the concept of "entrepreneurial discovery" in the context of smart specialization strategies, which is to define a policy mix and actions through a process of discovery and innovation driven by the engagement of companies, closer to "choosing races and placing bets" rather than "picking the winners" (Landabaso, 2014; McCann, 2015).

Most important is a change in the understanding of cities as the outcome of chaotic market transactions and coordinated, well-planned state interventions. This concept, which was a landmark of city planning throughout the Twentieth Century, is changing towards an understanding in which complex and chaotic forces operate on both the market and on the policy sides. The making of city plans comes closer to the concept of a laboratory of ideas and a roadmap of open innovation and entrepreneurship (Cohen et al., 2016) than meticulous elaboration and implementation of plans by central and local authorities. The notion of "planning without a plan" is about a smart city plan that is formed gradually, taking advantage of evolving technologies and opportunities for action. Both master plans and action plans of strategic planning are forms of top-down planning with well-defined plans and actions, in contrast to smart city planning which is shaped bottom-up, gradually, by user engagement and the capabilities offered by volatile technologies.

To better illustrate this understanding of city planning as an evolutionary process composed of urban laboratories and taking advantage of global and local opportunities, we consider it necessary to revise the roadmap of smart city planning that we presented in a previous publication (Komninos, Tsarchopoulos, and Kakderi, 2014). The ideas of "governance and feedback loops" between implementation, the focus on ecosystems, the selection of which challenges to address, and strategy development, which are added in Figure 1, express an open-minded management approach, which is non-linear, enabling cities to seize opportunities continuously and set up large-scale participation of citizens and organizations in city labs operating in various domains of the urban system.

The event horizon of this evolutionary smart city planning goes far beyond the physical space of cities, addressing all the grand challenges of twenty-first-century life in cities: the growth, employment, and poverty nexus; sustainability and its aspects, ranging from the use of land and nature-based solutions, to management of ecosystems, air quality, CO_2 emissions, climate adaptation, energy savings and the transition to renewable energy, water, waste recycling of materials, and the circular economy; and the urban safety nexus with man-made or natural threats, such as crime, terrorism, attacks on infrastructure, vandalism, natural catastrophes, urban accidents, and other types of emergencies. In sum, it addresses all aspects of cities, not just the physical space, land uses, and infrastructures addressed in nineteenth- and twentieth-century city planning.

Empowerment is the main pillar of strategy development, enabling intense information flows and knowledge sharing among users; easiness of collaboration; large-scale citizen engagement over crowdsourcing platforms; data creation, big datasets, and analytics; the rise of a sharing economy; few forms of production, such as demand-driven production, distributed collaborative production, customer co-production, and various other forms of network-based work and exchange.

Then, on the implementation and technology side, very competitive business models are based on open-source technologies, provided that they are carefully selected and supported by large and active communities of developers; cloud computing platforms, also developed with open source software, which disengage city authorities from technical and internal resource constraints; and open data initiatives offered via hackathons and competitions for the development of software and smart city solutions.

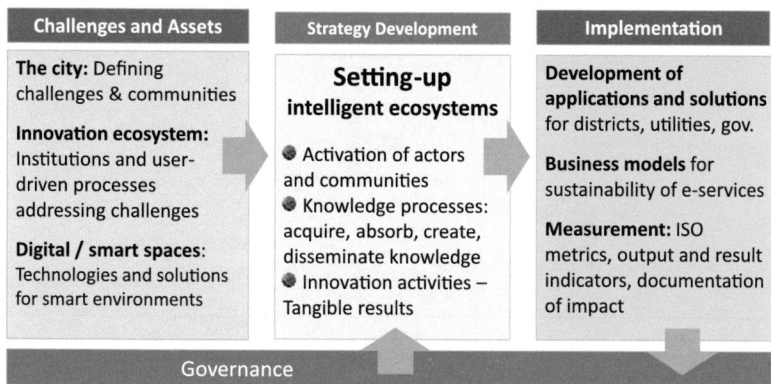

Figure 1. A roadmap for smart city planning. Source: Adapted from Komninos, Tsarchopoulos, and Kakderi, 2014.

Notes

1. The terms "smart city" and "intelligent city" are used interchangeably to mean the deployment of digital technologies, smart environments, and user engagement to optimize urban systems, and the economic and social life of cities. Some differences between these terms may be found in the way optimization takes place, as a direct outcome of technology or as an outcome of better decision-making. Thus, in our view, smart cities are related to solutions that optimize urban systems and user behavior through smart devices, ICT-based automation, sensors and instrumentation; while intelligent cities are related to solutions that enable people-driven innovation, improve decision-making through wider user engagement and datasets, advancing human intelligence and problem-solving capabilities (Komninos, 2014: 20–21).
2. STORM CLOUDS Project: See<http://storm-clouds.eu> Accessed May 28, 2017.
3. The European Innovation Partnership on Smart Cities and Communities<http://ec.europa.eu/eip/smartcities/> Accessed May 28, 2017,
4. Apps for Thessaloniki<http://thessaloniki.appsforgreece.eu> Accessed May 28, 2017.
5. Hackathon Thessaloniki<http://www.hackathess.eu> Accessed May 28, 2017.
6. Thessaloniki e-government portal<http://opengov.thessaloniki.gr> Accessed May 28, 2017.
7. IBM Smarter Cities Challenge – Thessaloniki<https://goo.gl/0CkA4i> Accessed May 28, 2017.
8. Open Data Portal of the Municipality of Thessaloniki<http://opendata.thessaloniki.gr> Accessed May 28, 2017.
9. Greek Cities Open Data Census<http://gr-city.census.okfn.org/> Accessed May 28, 2017.
10. "STORM CLOUDS: Surfing Towards the Opportunity of Real Migration to cloud-based public services" is a project co-funded by the CIP-ICT-PSP program of the European Commission.
11. 100 Resilient Cities<http://www.100resilientcities.org> Accessed May 28, 2017.
12. Thessaloniki Youth Resilience Challenge.
13. Project "PEOPLE: Pilot smart urban Ecosystems leveraging Open innovation for Promoting and enabLing future E- services" partly funded by the European Commission under contract No. 271027.

Disclosure Statement

No potential conflict of interest was reported by the authors.

ORCID

A. Panori ⓘ http://orcid.org/0000-0002-2551-2032
N. Komninos ⓘ http://orcid.org/0000-0002-4656-1263
C. Kakderi ⓘ http://orcid.org/0000-0001-6499-3919

Bibliography

M. Batty. *The New Science of Cities* (Cambridge: MIT Press, 2013).

B. Bergvall-Kåreborn and A. Ståhlbröst, "Living Lab: An Open and Citizen-Centric Approach for Innovation," *International Journal of Innovation and Regional Development* 1: 4 (2009) 356–370.

L. M. Bettencourt, J. Lobo, D. Strumsky, and G. B. West, "Urban Scaling and its Deviations: Revealing the Structure of Wealth, Innovation and Crime across Cities," *PloS one* 5: 11 (2010) e13541.

L. Bettencourt and G. West, "A Unified Theory of Urban Living," *Nature* 467: 7318 (2010) 912–913.

R. Boschma, "Competitiveness of Regions from an Evolutionary Perspective," *Regional Studies* 38: 9 (2004) 1001–1014.

A. Caragliu, C. Del Bo and P. Nijkamp, "Smart Cities in Europe," *Journal of Urban Technology* 18: 2 (2011) 65–82.

C. H. Chen-Ritzo, C. Harrison, J. Paraszczak, and F. Parr, "Instrumenting the Planet," *IBM Journal of Research* and *Development* 53: 3 (2009) 338–353.

B. Cohen, E. Almirall, and H. Chesbrough, "The City as a Lab: Open Innovation Meets the Collaborative Economy," *California Management Review* 59: 1 (2016) 5–13.

H. Couclelis, "The Construction of the Digital City," *Environment and Planning: Planning and Design* 31 (2004) 5–19.

M. Deakin, "Smart Cities and the Internet: From Mode 2 to Triple Helix Accounts of Their Evolution". In A. Vesco, ed., *Smart Cities Research Handbook: Social, Environmental and Economic Sustainability* (Hershey: IGI, 2015).

M. Deakin and S. Allwinkle, "Urban Regeneration and Sustainable Communities: The Role Networks, Innovation and Creativity in Building Successful Partnerships," *Journal of Urban Technology* 14: 1 (2007) pp. 77–91.

G. De Roo and E. A. Silva, *A Planner's Encounter with Complexity* (Farnham: Routledge, 2016).

S. Goldenberg, "Masdar's Zero-Carbon Dream Could become World's First Green Ghost Town", *The Guardian* (February 16, 2016) <https://www.theguardian.com/environment/2016/feb/16/masdars-zero-carbon-dream-could-become-worlds-first-green-ghost-town> Accessed on February, 16 2016.

P. Hall, *Cities of Tomorrow* (Blackwell, 1988).

K. Hiramatsu and T. Ishida, "An Augmented Web Space for Digital Cities" Symposium on Applications and the Internet Proceedings. (2001) <http://ieeexplore.ieee.org/xpl/freeabs_all.jsp?arnumber=905173> Accessed on February, 16 2016.

IBM *Thessaloniki, Greece 2017 Challenge*, Smarter Cities Challenge (2017) <https://www.smartercitieschallenge.org/cities/thessaloniki-greece> Accessed on December, 16 2017.

T. Ishida and K. Isbister, *Digital Cities: Technologies, Experiences, and Future Perspectives* (Berlin: Springer-Verlag, 2000).

C. Kakderi, N. Komninos, and P. Tsarchopoulos, "Smart Cities and Cloud Computing: Lessons from the STORM CLOUDS Experiment," *Journal of Smart Cities* 2: 1 (2016) 4–13.

N. Komninos, *Intelligent Cities and Globalisation of Innovation Networks* (London: Routledge, 2008).

N. Komninos, *The Age of Intelligent Cities: Smart Environments and Innovation-for-all Strategies* (New York City, NY: Routledge, 2014).

N. Komninos, "Smart Environments and Smart Growth: Connecting Innovation Strategies and Digital Growth Strategies", *International Journal of Knowledge-Based Development* 7: 3 (2016a), 240–263.

N. Komninos, "Intelligent Cities and the Evolution toward Technology-Enhanced, Global and User-Driven Territorial Systems of Innovation," *Handbook on the Geographies of Innovation* (2016b) 187–200.

N. Komninos and P. Tsarchopoulos, "Toward Intelligent Thessaloniki: From an Agglomeration of Apps to Smart Districts," *Journal of the Knowledge Economy* 4: 2 (2013) 149–168.

N. Komninos, P. Tsarchopoulos, and C. Kakderi, "New Services Design for Smart Cities: A Planning Roadmap for User-Driven Innovation," *Proceedings of the 2014 ACM International Workshop on Wireless and Mobile Technologies for Smart Cities* (2014) 29–38.

J. G. Lambooy, "Knowledge and Urban Economic Development: An Evolutionary Perspective," *Urban Studies* 39: 5-6 (2002) 1019–1035.

M. Landabaso, "Guest Editorial on Research and Innovation Strategies for Smart Specialisation in Europe: Theory and Practice of New Innovation Policy Approaches," *European Journal of Innovation Management* 17: 4 (2014) 378–389.

L. Leydesdorff and M. Deakin, "The Triple Helix of Smart Cities: A Neo-Evolutionist Perspective," *Journal of Urban Technology*, 18: 2 (2011) 53–64.

Living Plant, *Introduction to the PlanIT Urban Operating System™ Architecture* (2016) <http://living-planit.com/pdf/living-planit-introduction-to-uos-architecture-whitepaper.pdf> Accessed on May, 15 2017.

S. Marvin and A. Luque Ayala, "Urban Operating Systems: Diagramming the City," *International Journal of Urban and Regional Research* 41: 1 (2017) 84–103.

P. McCann, *The Regional and Urban Policy of The European Union: Cohesion, Results-Orientation and Smart Specialisation* (Cheltenham: Edward Elgar, 2015).

Municipality of Thessaloniki, *Digital Strategy 2017–2030* (Thessaloniki: 2017) <https://opengov.thessaloniki.gr/nea/108-parousiasi-tis-psifiakis-stratigikis-tou-dimou-thessalonikis> Accessed on November, 5 2017.

R. R. Nelson and S. G. Winter, "In Search of Useful Theory of Innovation," *Research Policy* 6: 1 (1977): 36–76.

T. O'Reilly, "What Is Web 2.0: Design Patterns and Business Models for the Next Generation of Software," *International Journal of Digital Economics* 65 (2007) 17–37.

N. Rosenberg, *Inside the Black Box: Technology and Economics* (Massachusetts: Cambridge University Press, 1982).

J. Simmie and R. Martin, "The Economic Resilience of Regions: To wards an Evolutionary Approach," *Cambridge Journal of Regions, Economy and Society* 3: 1 (2010) 27–43.

N. Streitz, "Reconciling Humans and Technology: The Role of Ambient Intelligence," *European Conference on Ambient Intelligence* (2017) 1–16.

Thessaloniki, *Resilient Thessaloniki*, 100 Resilient Cities (2017) <www.100resilientcities.org/strategies/city/thessaloniki#/-_/> Accessed on 17 November 2017.

I. Tsampoulatidis, D. Ververidis, P. Tsarchopoulos, S. Nikolopoulos, I. Kompatsiaris, and N. Komninos, "Improvemycity: An Open Source Platform for Direct Citizen-Government Communication," *Proceedings of the 21st ACM International Conference on Multimedia* (2013) 839–842.

P. van den Besselaar and S. Koizumi, *Digital Cities III. Information Technologies for Social Capital: Cross-cultural Perspectives* (Berlin: Springer, 2005).

G. S. Yovanof and G. N. Hazapis, "An Architectural Framework and Enabling Wireless Technologies for Digital Cities & Intelligent Urban Environments," *Wireless Personal Communications* 49: 3 (2009) 445–463.

T. Yigitcanlar, *Technology and the City: Systems, Applications and Implications* (London: Routledge, 2016).

Smart Cities and Mobility: Does the Smartness of Australian Cities Lead to Sustainable Commuting Patterns?

Tan Yigitcanlar ⓘ and Md. Kamruzzaman ⓘ

ABSTRACT

Smart cities have become a popular concept because they have the potential to create a sustainable and livable urban future. Smart mobility forms an integral part of the smart city agenda. This paper investigates "smart mobility" from the angle of sustainable commuting practices in the context of smart cities. This paper studies a multivariate multiple regression model within a panel data framework and examines whether increasing access to broadband Internet connections leads to the choice of a sustainable commuting mode in Australian local government areas. In this case, access to the Internet is used as a proxy for determining urban smartness, and the use of different modes of transport including working at home is used to investigate sustainability in commuting behavior. The findings show that an increasing access to broadband Internet reduces the level of working from home, public transport use, and active transport use, but increases the use of private vehicles, perhaps to overcome the fragmentation of work activities the Internet creates. How to overcome the need for car-based travel for fragmented work activities while increasing smartness through the provisioning of broadband access should be a key smart city agenda for Australia to make its cities more sustainable.

Introduction

The concept, "smart city," has become almost ubiquitous both in academia and in policy circles due to its potential to address a range of negative effects of rapid urbanization (e.g., congestion, CO_2 emissions), industrialization (e.g., air and soil pollution) and consumerism practices (Mahbub et al., 2011; Wiig, 2015; Taamallah et al., 2017; Trindade et al., 2017). While a number of scholars and smart city sceptics raise their concerns about the ongoing global smart city movement (Yigitcanlar and Lee, 2014; Kunzmann, 2015; Angelidou, 2017), many levels of government—local, regional, state, national, and supra national—across the globe still continue to jump on the smart cities bandwagon (Townsend, 2013; Komninos, 2016). Due to diverse disciplinary and sectoral perspectives, there is no common consensus on the definition of smart cities (Angelidou, 2014; Albino et al., 2015). However, these cities are generally seen as localities that effectively utilize strategic planning approaches and innovative solutions to improve the quality of life of their communities, including ecological, cultural, political, institutional, social, and economic components (Neirotti et al., 2014; Yigitcanlar, 2016). Smart cities are also an umbrella concept that contain various sub-elements, ranging from smart economy to smart living, smart governance to smart people, and smart mobility to smart environment (Lee et al., 2014; Lara et al., 2016; Chang et al., 2018).

Because emissions generated from transport causes about a quarter to one-third of greenhouse gas (GHG) emissions, smart mobility forms an integral part of the smart city agenda (Creutzig et al., 2015; Yigitcanlar, 2015; Arbolino et al., 2017). The smart mobility concept, in the fashionable sense, is defined as "integrating the sustainable and smart vehicular technologies, and the cooperative-ITS [intelligent transport systems], accelerated with the cloud-server and big-data based vehicular networks" (Kim et al., 2015, p.59). Similarly, Chun and Lee (2015) see smart mobility as a comprehensive and smarter future traffic service in combination with smart technology. However, in the traditional sense, smart mobility is basically all about reducing congestion, greenhouse gases, and other vehicular emissions, and fostering faster, greener, and cheaper transportation options (Spinney et al., 2009). Moving smartly surely depends on an efficient means of active and public transport systems having a low environmental impact, a network of safe and continuous cycle lanes and walkways, and interchange parking that avoids city congestion (Yigitcanlar and Kamruzzaman, 2014, 2015; Chun and Lee, 2015). In other words, mobility cannot be considered "smart" if it is not also sustainable (Yigitcanlar et al., 2015; Garau et al., 2016).

This study aims to capture the big picture view on the relationship between urban smartness, measured in terms of access to the Internet, and sustainable forms of commuting. In line with this aim, the study focuses on addressing the research question: Do the residents in smarter cities commute more sustainably? In order to address this critical question, the paper concentrates on the following research objectives: (a) Establishing a causal link between urban smartness and new forms of working such as working from home, and (b) Exploring whether changes in urban smartness alter inhabitants' choice of commuting mode.

The methodological approach of this research includes a thorough review of the literature and applying multivariate multiple regression analysis to address the abovementioned research question and execute the research objectives. The case study cities for the empirical investigation are selected from Australia—all local government areas. The selection of Australia as the study context is justified because of (a) the Federal government's recent interest and investment in the smart cities agenda (Australian Government, 2016); (b) Australian cities' success (and failure) experiences in becoming smart cities (Yigitcanlar, 2016); (c) Ever-continuing political and public debates on the appropriate formulation of the delivery of a nationwide fast and reliable broadband network (Glance, 2017).

Literature Review

A decade into the commencement of the contemporary conceptualization and practice of the smart cities notion, the concept is still in its infancy (Alizadeh, 2017; Yigitcanlar, 2017; Praharaj et al., 2018). Today, many scholars are advocating the importance of urban planners and decision-makers being prepared for the onslaught of disruptive urban technologies—whether it is Internet of Things (IoT), social robotics, sharing economy, big data, artificial intelligence, crowdsourcing, drones, or 3D printing (Batty et al., 2012; Batty, 2013; Anthopoulos, 2017). Similarly, smart mobility—although under the ITS brand dating back to the 1980s—is also a relatively new brand, and it has the potential to bring out both the best or worst in our cities by transforming among other things, land use and mobility patterns; for example, as a consequence of the adoption of autonomous vehicles (Firnkorn and Müller, 2015; Truong et al., 2017). Instead of solely focusing on these two fashionable concepts, this literature review concentrates on their more traditional cores—the interplay between information and communication technologies (ICTs) and transport systems, along with their combined impacts on sustainable commuting.

The research on the impacts of ICTs on transport systems has matured over the last four decades (Goldmark, 1969; Messenger and Gschwind, 2016). Given that both transport and ICT are considered communication technologies, the initial expectations were that the old transport technology would be replaced by new ICT technology because of the reduced generalized cost of reaching amenities and services and the improved quality and attractiveness of those amenities and services (Lyons, 2002). Such replacement, as expected, would have three types of impacts: (a) changes in travel demand; (b) changes in transportation systems such as the development of ITS and consequent efficiency gains; (c) changes in urban form caused by the demand for certain types of land uses and the accessibility of such places (Brown et al., 2005; Cohen-Blankshtain and Rotem-Mindali, 2016). This review focuses on the impacts of ICT on travel demand, and more specifically, on travel behavior.

The presumption that ICT potentially changes travel behavior stems from a hypothetical understanding that ICT will replace aspects of the traditional transport system and support sustainable mobility. Numerous studies have been conducted to prove empirically this hypothesis and several review articles have also been published based on these empirical studies (Aguiléra et al., 2012; Cohen-Blankshtain and Rotem-Mindali, 2016). However, most of these studies have focused on investigating the links between telecommunications and travel behavior (Mokhtarian, 1991; Claisse and Rowe, 1993; Fadare and Salami, 2004; Kwan et al., 2007). These studies have shown that telecommunications have four types of effects on travel behavior: (a) substitution (e.g., telecommunications replace travel); (b) enhancement/complementarity (e.g., telecommunications increase travel); (c) modification (e.g., telecommunications change the way people travel); (d) neutrality (e.g., telecommunications have no effect on travel) (Brown et al., 2005; Zhang et al., 2007).

Transport researchers have started examining the impacts of the Internet on the abovementioned four dimensions of travel in the last decade. These studies have broadly been classified into two groups: (a) general use of the Internet and activity-travel patterns; (b) specific use of the Internet (e.g., teleshopping, telecommuting) and their impacts on overall travel (Ren and Kwan, 2009). A range of indicators in relation to the general use of the Internet have been used in the literature such as the frequency and/or amount of Internet use (Viswanathan and Goulias, 2001; Nobis and Lenz, 2004; Srinivasan and Reddy, 2004; Zhang et al., 2007; Kenyon, 2010); and the use of home computers stratified by: (a) private tasks without need of an Internet connection; (b) private tasks with need of an Internet connection; (c) work tasks without need of an Internet connection;

(d) work tasks with need of an Internet connection (Hjorthol, 2002). Typical variables used to assess the impacts of the Internet include vehicle kilometers traveled, total daily trips, daily walking trips, mode of travel, and time spent traveling (Viswanathan and Goulias, 2001; Hjorthol, 2002; Nobis and Lenz, 2004; Zhang et al., 2007).

Mostly, the findings from the studies on the impacts of the Internet on travel behavior correspond to the four types of telecommunication effects as outlined previously (i.e., substitution, complementarity, modification, neutrality). For example, a few studies have reported that ICT use is proportionately related with: (a) trip making (Johansson, 1999; Wang and Law, 2007; Zhang et al., 2007; Lila and Anjaneyulu, 2016); (b) car-dependency (Nobis and Lenz, 2004); (c) travel distance (Hjorthol, 2002; Zhang et al., 2007); and (d) travel time (Wang and Law, 2007). The main reason for the complementarity effect of the Internet on travel behavior is the result of widened travel horizons, an increase in time available for travel, the productive use of travel time, the intrinsic value of travel as an activity in itself and the ability of the Internet to make travel itself more effective (Kenyon, 2010). Other studies found the evidence of the replacement effect of the Internet on travel behavior. For example, Viswanathan and Goulias (2001) revealed that Internet use was correlated negatively with time spent on travel.

The relationship between Internet use and travel behavior is not always straightforward, but possesses many complexities depending on the types of users and usage of the Internet as well as the types of activities investigated. For instance, Hjorthol (2002) indicated that the users of home computers tend to make somewhat fewer work trips but do more chauffeuring and total trips. Similarly, Handy and Yantis (1997) found that in-home entertainment activities generate additional travel whereas online maintenance activities reduced travel. The impacts of Internet activities on travel also vary with gender. For example, Ren and Kwan (2009) revealed that Internet use for maintenance purposes has a greater impact on women's activity-travel in the physical world, while Internet use for leisure purposes affects men's physical activities and travel to a greater extent. Therefore, the results indicate the presence of both substitution and generation effects due to Internet use (Srinivasan and Reddy, 2004).

The Internet can modify travel behavior in a number of ways. One of the ways found in the literature is the fragmentation effect (Couclelis, 2000; Schwanen and Kwan, 2008)—i.e., separation of activities into discrete pieces (e.g., decomposition of work into multiple segments of subtasks, which can be performed at different times and/or locations). Alexander et al. (2010) found that Internet use can lead to three types of fragmentation of work activities: (a) a less temporally and spatially fragmented work pattern; (b) a less spatially but more temporally fragmented work pattern; (c) a more spatially and temporally fragmented work pattern. Fragmentation often leads to an increase in travel as activities are no longer tied to particular places and/or times (Kwan et al., 2007).

Recent studies also reported the neutrality effect of the Internet. For example, Line et al. (2011) found that the effect of the Internet on changes in social practice at the level of the individual is not visibly dramatic among students and part-time working mums. However, they reported that such technologies enable them to better accommodate the uncertainties in activity and travel scheduling—reflecting the modification effect.

Studies focusing on the specific use of the Internet also found mixed results. These studies investigated three types of specific use of the Internet such as teleworking (or telecommuting), teleshopping, and tele-leisure, and empirically tested their effects on travel behavior. Telecommuting occurs when commuting trips are replaced by virtual commuting as individuals are able to work in their homes (Aguiléra et al., 2012). Further refinement of telecommuting is done in the literature into three modes of work: home office, mobile office, and virtual office (Messenger and Gschwind, 2016). A number of authors investigated the effects of teleshopping on travel behavior. They found that teleshopping affects travel behavior in a number of ways, including replacing travel when online shopping is delivered at home (Anderson et al., 2003; Weltevreden, 2007) and reducing unnecessary travel (e.g., visiting multiple shops to explore bargains and offers). Other studies, however, reported no/limited effects of e-commerce when shopping trips are combined with other activities (e.g., work), or trips are made anyway for the purchase of other goods (Mokhtarian, 2004). It is also evident that teleshopping has the complementary effect. For example, increasing advertisement of retailers for bargains may induce additional trips or travel to shops further away that were not previously known to the customers (Rotem-Mindali, 2010).

Scholars have used a range of datasets to investigate the effects of the Internet on travel behavior including (a) nationally representative samples such as the National Household Travel Survey data in the United States (Zhang et al., 2007) and the Norwegian National Personal Travel Survey data (Hjorthol, 2002); (b) travel-diary data from representative samples in Germany (Nobis and Lenz, 2004) and the San Francisco Bay Area (Srinivasan and Reddy, 2004); (c) specifically designed survey data for the purpose of identifying Internet use and travel in Sweden (Johansson, 1999); and (d) qualitative diary and interview data (Line et al., 2011). A common aspect of all these datasets is that they are cross-sectional in nature and the analysis presented provides a snapshot of Internet use and travel behavior in a point in time. These scholars applied a range of analytical techniques to estimate the effect of Internet use on travel behavior such as linear regression models (Hjorthol, 2002; Zhang et al., 2007), the Poisson regression model to investigate the variations in the number of trips made (Zhang et al., 2007), and more complex structural equation models (Wang and Law, 2007; Ren and Kwan, 2009; Lila and Anjaneyulu, 2016). However, given the cross-sectional nature of the data used, their analyses lack the ability to ascertain causal relationships between Internet use and travel behavior (Zhang et al., 2007).

Studies highlighted that a causal relationship is more certain when a change in explanatory factor (Internet use in this case) causes a change in travel behavior (Handy et al., 2005; Kamruzzaman et al., 2016). In this regard, Kenyon (2010) suggested that panel-based research is more likely than cross-sectional research to accurately uncover the strength and form of relationships. Despite this, Kenyon (2010) collected longitudinal panel data, but applied a cross-sectional analysis method. The current evidence base is, therefore, insufficient to answer the research question raised earlier. Clearly, there is a need to expand the empirical investigation base with cause-effect mechanisms to clearly understand the connections between technological changes in cities and their sustainability outcome.

Empirical Investigation

Case Study

There were several reasons for conducting the empirical study in Australia. First, after the Australian Prime Minister's 2015 announcement highlighting cities as a national priority, the Commonwealth's Smart Cities Plan was launched in 2016. The plan aims to deliver jobs closer to homes, more affordable housing, better transport connections, and healthy environments. It includes: (a) establishing of an infrastructure financing unit to work closely with the private sector on innovative financing solutions and (b) committing funding to accelerate planning and development works on major infrastructure projects to develop business cases and investment options (Australian Government, 2016). Although the smart cities agenda only dates back to 2015, Australia has long been a nation that enjoys much of what technology has to offer. For example, significant clusters of wired communities started to become evident in Sydney as early as 2001 (Baum et al., 2004). Similarly, around the same time, most of the Australian local governments started to use ICTs in their daily urban and transport planning tasks (Yigitcanlar, 2005, 2006).

Second, Australian cities have a high-level of vulnerability to the likely consequences of global climate change as they are located in the world's driest continent (Goonetilleke et al., 2014). Against this challenge, a number of cities in Australia showcase various success (and also some failure) stories in their journey to become smart (and thus sustainable) cities. Sydney, Melbourne, Brisbane, Perth and Adelaide are among the cities with strategies in place and are progressing firmly on their smart city agendas. For instance, Brisbane's "City Smart" strategy includes: (a) creating a legible structure plan, (b) uniting disparate precincts, (c) creating definitive pedestrian spines, (d) linking the city center by mass transit, (e) defining a knowledge corridor, (f) investing in sustainability, (g) developing effective planning processes, and (h) developing a smart city model (Hortz, 2016; Yigitcanlar, 2016).

Finally, Australia is behind most of the other OECD countries in terms of a fast and reliable broadband Internet infrastructure. For example, in 2013, Australia was ranked 29 out of 34 OECD countries in average Internet connection speed (OECD, 2014). One of the reasons for this is that political, industry, and public debates lasted for too long and delayed the commencement of the development and thus the completion of the national broadband network (NBN) project—offering up to 100 Mbps download and 40 Mbps upload speeds (Yigitcanlar, 2016). The geographic coverage of the NBN is limited, as this service is mostly available to small parts of the metropolitan cities, and the rest of the country is generally served with ADSL2+—at best up to 24 Mbps download and 1.4 Mbps upload speeds (Yigitcanlar, 2016). Moreover, there are high numbers of connection delay complaints (Glance, 2017). According to Tucker and Branch (2013), at the current rate of NBN roll-out, the project may take more than two decades to cover the whole country. This would not only jeopardize the smart cities agenda in the country, but also cause a major disadvantage in Australia's competitiveness in the digital economy (Bowles and Wilson, 2012).

A country context with aforementioned challenges and opportunities make Australia an interesting testbed for research that investigates whether the smartness of cities lead to sustainable commuting patterns.

Data

This research used the 2006 and 2011 editions of the census data collected by the Australian Bureau of Statistics (ABS). Data are collected from all people (including visitors) on a specified census day in Australia every five years. This is the only source of data that provides information for the entire country, also stratified by different administrative boundaries. This research used Local Government Area (LGA) boundary as the unit of analysis because Australian Commonwealth (Federal) Government funds, to promote smart city agenda in Australia, are channeled through the LGAs (https://cities.dpmc. gov.au/smart-cities-plan). Note that there were changes in the number of LGAs in Australia between 2006 (n = 676) and 2011 (n = 568). For example, some LGAs were merged together or the names of some of the LGAs have been slightly modified. This research used data only from those LGAs that remained fixed in both periods, which resulted in 513 analytical samples of LGAs.

The main exposure variable of interest is access to the Internet. In both the 2006 and 2011 editions of the survey, questions were asked to respondents whether their dwelling is connected with the Internet, and if yes, the type of connection (e.g., broadband, dial-up, others). The datasets also provided information about the respondents who did not answer this question and information about those for whom this question was not applicable (e.g., not living in a private dwelling). Note that the 2016 edition of the survey contained a modified version of the question and asked whether private dwellings have any people who access the Internet from the dwelling, without details of the type of connection. Given that the main objective of this research is to test the causal relationships between access to the Internet and commuting behavior, and that this required a panel nature of longitudinal data, the 2016 edition of the survey data was not included because the questions were not consistent with those of the prior surveys. With the interest of modelling the impact of changes, this research required calculating the percentage of dwellings with Internet connections (and types) in both periods because LGAs experienced population/dwelling growth over the period. As a result, relative changes were derived (percentage of changes in dwellings with Internet connections) rather than absolute changes (changes in the number of dwellings).

The commuting mode of transport was used as the outcome variable of this research. In both surveys, respondents aged 15+ were asked to indicate their "method of travel to work." They were given 14 options to choose from: train, bus, ferry, tram (including light rail), taxi, car-as driver, car-as passenger, truck, motorbike or motor scooter, bicycle, walked only, worked at home, other, and did not go to work. Respondents were also given the option of choosing multiple modes of transport. This research reclassified the modes of transport into the following categories: public transport (merging train, bus, ferry, and tram), private transport (merging taxi, car-as driver, car-as passenger, truck, motorbike or motor scooter), active transport (merging bicycle, walked only), other, worked at home, did not go to work, and not stated (respondents who did not answer to this question). In cases where respondents selected multiple modes, these were assigned to the above categories according to the following priorities: public transport (if public transport was a selected mode among the multiple modes), private transport (if private transport was a selected mode among the multiple modes but not public transport), and active transport (if active transport was a selected mode among the multiple modes but not public transport/private transport). Like the access-to-the-Internet variable, relative changes between the periods in these modal categories were derived. Table 1 outlines descriptive statistics for both exposure and outcome variables.

As shown in Table 1, this research also derived a range of other variables, including household income, vehicle ownership, employment status, and gender. These variables

Table 1. Variable description

Variables	Mean	Std. Dev.	Min	Max
Outcome variables (changes in employed person aged 15+ between 2006 and 2011)				
% of changes in persons worked at home	−1.10	2.02	−14.45	10.56
% of changes in persons commuting by public transport	0.83	2.14	−8.35	17.22
% of changes in persons commuting by private transport	2.02	4.41	−28.38	28.70
% of changes in persons commuting by active transport	−1.23	3.96	−22.35	39.82
% of changes in persons commuting by other mode of transport	0.12	2.23	−42.15	12.26
% of changes in persons did not go to work	−0.32	2.23	−9.33	32.77
% of changes in persons not stating commuting mode	−0.31	1.37	−8.93	16.67
Explanatory variables				
Change factors between 2006 and 2011				
Internet connection: % of changes in dwellings with				
Broadband connection	35.89	9.27	0.00	62.14
Dial-up connection	−16.21	6.34	−42.07	0.12
Other types of Internet connection	3.75	3.41	0.00	65.89
No Internet connection	−8.95	6.48	−66.34	26.08
Internet connection not applicable	−15.14	9.28	−61.23	0.00
Internet connection not stated	0.64	2.89	−20.82	13.46
Income: % of changes in families with income (per week)				
Negative or zero income	0.41	1.17	−7.92	5.93
$1–$999	10.94	8.51	−76.47	37.33
$1k–$1,999	−4.60	8.48	−31.34	76.47
$2k–$2,999	−6.94	4.10	−30.00	8.33
$3k–$3,999	−0.84	1.98	−10.97	12.79
$4k or more	1.02	1.78	−7.09	19.94
Vehicle ownership: % of changes in dwellings with				
Zero vehicle	−0.73	1.79	−20.70	13.40
One vehicle	−0.37	2.63	−22.66	12.71
Two vehicles	−0.22	2.27	−10.94	8.60
Three vehicles	0.53	1.57	−15.79	8.62
Four or more vehicles	0.73	1.22	−4.29	9.21
Vehicles not stated	−0.86	2.66	−15.79	9.67
Vehicles not applicable	0.92	3.67	−17.86	38.89
Gender: % of changes in persons described as				
Female	−0.15	2.39	−16.09	28.54
Employment status: % of changes in employed persons described as				
Full-time employment	−0.02	5.88	−19.66	45.41
Part-time employment	0.02	5.88	−45.41	19.66
Base factors in 2006				
Commuting: % of employed person				
Worked at home	9.57	7.26	0.00	37.54
Commuted by public transport	4.42	6.70	0.00	34.82
Commuted by private transport	60.99	13.65	4.00	80.92
Commuted by active transport	11.00	12.15	1.06	91.72
Commuted by other mode	1.61	3.82	0.00	63.16
Did not go to work	10.34	2.76	0.00	17.43
Not Stated commuting mode	2.08	1.15	0.00	12.12
Internet connection: % of dwellings with				
Broadband connection	22.84	12.02	0.00	62.15
Dial-up connection	19.39	7.05	0.00	48.33
Other types of Internet connection	0.51	0.32	0.00	2.15
No Internet connection	34.23	10.01	0.00	80.00
Internet connection not applicable	16.60	11.68	0.00	101.43
Internet connection not stated	6.42	3.90	0.00	41.51
Income: % of families with income				
Negative or zero income	1.31	1.32	0.00	9.52
$1–$999	43.80	13.50	5.67	100.00
$1k–$1,999	37.77	7.14	0.00	68.42
$2k–$2,999	12.04	6.88	0.00	41.96
$3k–$3,999	3.47	4.10	0.00	20.42

(Continued)

Table 1. Continued.

Variables	Mean	Std. Dev.	Min	Max
$4k or more	1.61	2.42	0.00	24.51
Vehicle ownership: % of dwellings with				
Zero vehicle	7.83	7.48	0.00	55.20
One vehicle	27.73	6.51	0.00	47.54
Two vehicles	27.35	7.31	0.00	43.52
Three vehicles	8.92	3.29	0.00	20.31
Four or more vehicles	4.86	2.85	0.00	19.58
Vehicles not stated	6.67	3.72	0.00	28.57
Vehicles not applicable	16.64	11.70	0.00	100.00
Gender: % of persons described as				
Female	49.14	3.90	7.67	56.87
Employment status: % of employed persons described as				
Full-time employment	68.22	7.88	22.73	96.25
Part-time employment	31.78	7.88	3.75	77.27

N = 513 (local government areas)

have commonly been identified to have significant effect on the choice of commuting mode (Asensio, 2002; Beckman and Goulias, 2008; Antipova et al., 2011; Kamruzzaman et al., 2014; Cao, 2015; Clark et al., 2016). As a result, changes in these variables between the periods were used as controlling factors to investigate the causal link between access to the Internet and commuting travel behavior. Other time invariant factors such as the remoteness index of the LGAs were not considered because of the panel nature of the data analysis technique applied in this research which differenced out such factors.

Approach: Multivariate Multiple Regression Analysis

The way variables were coded (as described above) in this research can be regarded as panel data with two-time periods for the 513 LGAs. When variables are observed at only two periods, an ordinary least square (OLS) regression model can reliably generate unbiased estimates of coefficients on the difference scores between the time periods (Allison, 2009). However, an OLS model is suitable to estimate a coefficient only when there is a single outcome variable. This research requires modeling five outcome variables (two outcome variables—did not go to work, and commuting mode not stated—were not included in the analysis due to their lack of relevance) with moderate levels of correlations among some of the outcome variables (See Table 2). The observed correlations are expected in this research because the differences in mode choice behavior are calculated in percent, and as a result, if the use of one mode increases, naturally the use of other

Table 2. Correlations among the outcome variables

Outcome variables	% of changes in				
	Worked at home	Public transport use	Private transport use	Active transport use	Other mode use
Worked at home	1				
Public transport use	−0.1684	1			
Private transport use	−0.3127	−0.2975	1		
Active transport use	0.0594	−0.098	−0.7018	1	
Other mode use	−0.1517	0.11	−0.2167	0.0158	1

modes decreases. Research highlighted that when there are multiple outcome variables and at least a moderate level of correlation exists among the outcome variables, multivariate regression is a better suited model over the OLS model (Washington et al., 2010; Kamruzzaman and Hine, 2013).

In a multivariate model, the outcome variables are simultaneously regressed against the explanatory variables. When there is a single explanatory factor in a multivariate regression, it is called multivariate (simple) regression whereas a multivariate regression model is referred to as multivariable multiple regression when there is more than one explanatory factor in a model. More specifically, a multivariate multiple regression is "multiple" because there is more than one independent variable. It is "multivariate" because there is more than one dependent/outcome variable. Interested readers are referred to Dattalo (2013) for a detailed discussion on multivariate multiple regression models. In summary, given the need to model multiple outcome variables with moderate correlations which need to be regressed by more than one predictor variable, as a result, multivariate multiple regression models were estimated as the main analytical method in this research in order to examine whether changes in access to the Internet affected commuting patterns over the period of 2006–2011. The outputs of a multivariate regression model are interpreted in the same way as outputs from an OLS regression model.

This research models the changes in outcome variables (model of commuting choice) in an effort to investigate the impacts of changes in access to the Internet. Research has shown that individuals' changed behavior is a function of not only changed circumstances (e.g., changes in access to the Internet) but also related to their "base" values (status of access to the Internet in 2006) including choice of commuting mode in the base year of 2006 (Krizek, 2003; Kamruzzaman et al., 2016). As a result, base values associated with the choice of commuting mode, socio-demographic status of the LGAs, and status of the Internet connection in 2006 were used as explanatory factors in addition to the changed factors (e.g., changes in access to the Internet and socio-demographic status). This procedure is considered as the best way to define changes by correcting for the phenomenon of regression to the mean (Twisk, 2003).

The number of explanatory factors, however, became very large with the inclusion of both base and changed factors. (See Table 3.) This may result in an over-specification of the multivariate model given the 513 data points used in this research (Wilson et al., 2006). Model over-specification typically results in producing numerically unstable estimates and is characterized by unrealistically large estimated coefficients and/or estimated standard errors (Bursac et al., 2008). In addition, modelling exercises often seek to build a parsimonious model by minimizing the number of variables but without compromising the true outcome experience of the data. Hosmer et al. (2013: 90) stated that "the rationale for minimizing the number of variables in the model is that the resultant model is more likely to be numerically stable and is more easily adopted for use." Several methods exist in the literature to overcome the over-specification problem in a model (see Wilson et al., 2006; Bursac et al., 2008).

This research applied four different strategies to overcoming this problem. First, variables that are undefined (e.g., not stated or not applicable) were not included. Second, the research tested the correlations among the explanatory factors and highly correlated factors were excluded from further analysis in order to avoid multicollinearity problem. Table 3 lists a subset of the explanatory factors with high correlation coefficients. Third,

Table 3. Correlations among a few explanatory factors[a]

Explanatory factors	1	2	3	4	5	6	7	8	9	10	11	12	13	14	15	16	17	18	19	20	21	22	23
1. Changes in no Internet connection	1.00																						
2. Changes in dial-up connection	0.15	1.00																					
3. Changes in broadband connection	0.00	−0.79	1.00																				
4. Changes in not applicable (not residential) Internet connection	−0.69	−0.05	−0.41	1.00																			
5. % of no connection 2006	−0.11	0.18	−0.32	0.21	1.00																		
6. % of broadband connection 2006	−0.39	0.22	−0.36	0.56	−0.53	1.00																	
7. % of dialup connection 2006	−0.14	−0.99	0.83	0.00	−0.19	−0.24	1.00																
8. % of not applicable (not residential) Internet connection 2006	0.70	0.10	0.34	−0.98	−0.19	−0.58	−0.06	1.00															
9. % of not stated connection 2006	−0.02	0.36	−0.38	0.20	−0.11	0.13	−0.37	−0.15	1.00														
10. Changes in $1k−$1,999 income	0.03	0.33	−0.47	0.26	−0.01	0.34	−0.37	−0.26	0.26	1.00													
11. Changes in $1−$999 income	−0.12	−0.31	0.41	−0.12	−0.28	0.05	0.34	0.08	−0.23	−0.80	1.00												
12. % of private transport used 2006	−0.44	−0.28	0.26	0.29	−0.18	0.35	0.30	−0.38	−0.19	−0.20	0.28	1.00											
13. % of active transport used 2006	0.44	0.41	−0.47	−0.17	0.52	−0.44	−0.44	0.25	0.18	0.30	−0.49	−0.81	1.00										
14. Changes in $2k−$2,999 income	0.17	−0.06	0.14	−0.28	0.55	−0.65	0.09	0.29	−0.24	−0.32	−0.23	−0.17	0.30	1.00									
15. Changes in $3k−$3,999 income	0.16	−0.08	0.15	−0.26	0.36	−0.59	0.09	0.32	−0.02	−0.36	−0.17	−0.16	0.27	0.61	1.00								
16. Changes in $4k or more income	−0.10	0.15	−0.22	0.20	−0.24	0.29	−0.17	−0.12	0.32	0.24	−0.34	0.08	−0.02	−0.26	0.14	1.00							
17. % of negative/zero income 2006	0.16	−0.11	0.20	−0.20	−0.16	−0.12	0.11	0.24	−0.08	−0.13	0.14	−0.34	0.08	0.08	0.17	−0.03	1.00						
18. % of $1−$999 income 2006	0.28	0.05	0.02	−0.30	0.70	−0.75	−0.03	0.30	−0.18	−0.10	−0.33	−0.23	0.45	0.81	0.54	−0.32	−0.02	1.00					

(Continued)

Table 3. Continued.

Explanatory factors	1	2	3	4	5	6	7	8	9	10	11	12	13	14	15	16	17	18	19	20	21	22	23
19. % of $2k–$2,999 income 2006	-0.22	0.12	-0.23	0.35	-0.60	0.77	-0.15	-0.35	0.28	0.46	0.02	0.17	-0.29	-0.92	-0.62	0.44	-0.13	-0.87	1.00				
20. % of $3k–$3,999 income 2006	-0.14	0.19	-0.27	0.30	-0.56	0.75	-0.21	-0.31	0.24	0.52	-0.06	0.04	-0.21	-0.73	-0.77	0.32	-0.05	-0.76	0.86	1.00			
21. % zero vehicles 2006	0.05	0.48	-0.66	0.23	0.59	-0.08	-0.53	-0.19	0.16	0.51	-0.56	-0.47	0.71	0.15	-0.04	0.01	-0.11	0.33	-0.09	0.03	1.00		
22. % two vehicles 2006	-0.48	-0.53	0.35	0.42	-0.31	0.46	0.54	-0.48	-0.34	-0.23	0.39	0.74	-0.77	-0.25	-0.20	0.03	-0.12	-0.42	0.25	0.15	-0.61	1.00	
23. % three vehicles 2006	-0.21	-0.67	0.52	0.15	-0.21	0.06	0.68	-0.20	-0.37	-0.27	0.25	0.43	-0.51	0.01	0.11	-0.04	0.07	-0.19	-0.02	-0.12	-0.64	0.75	1.00
24. % 4+ vehicles 2006	-0.01	-0.63	0.55	-0.08	-0.18	-0.17	0.64	0.06	-0.30	-0.39	0.34	0.00	-0.24	0.08	0.21	-0.16	0.30	-0.10	-0.15	-0.23	-0.54	0.39	0.73
25. % not stated vehicles 2006	-0.02	0.37	-0.39	0.21	-0.12	0.16	-0.39	-0.16	0.97	0.28	-0.21	-0.19	0.18	-0.30	-0.07	0.31	-0.11	-0.21	0.33	0.27	0.15	-0.34	-0.37
26. % not applicable (not residential) vehicles 2006	0.70	0.10	0.34	-0.98	-0.19	-0.58	-0.06	1.00	-0.15	-0.26	0.08	-0.38	0.26	0.29	0.32	-0.12	0.24	0.30	-0.35	-0.31	-0.19	-0.48	-0.20
27. % active transport use 2006	0.44	0.41	-0.47	-0.17	0.52	-0.44	-0.44	0.25	0.18	0.30	-0.49	-0.81	1.00	0.30	0.27	-0.02	0.08	0.45	-0.29	-0.21	0.71	-0.77	-0.51

[a] Please note that not all independent factors, as presented in Table 1, are included in the table due to space restriction. It shows some factors with potentially high levels of correlations.

the research applied the purposeful selection method as laid out by Hosmer et al. (2013). Briefly, a multivariable (simple) regression model was estimated separately for each of the independent variables on the outcomes to identify factors that have a significant association with the outcomes. Only factors that were found to be significant at the $p < 0.1$ level were entered (forced entry) into the final multivariate multiple regression model (Bursac et al., 2008). Fourth, the multicollinearity among the selected explanatory factors was tested using ordinary least squares (OLS) regression model. A similar technique has been applied in previous research (Piya et al., 2013). The variance inflation factor (VIF) test showed that several of the significant factors (percentage of households with broadband connection in 2006, percentage of households with zero vehicles in 2006) suffer from multicollinearity problems with greater than 10 VIF. These variables were gradually removed and the OLS model was rerun until the problem was resolved.

The refined sets of explanatory factors were then used to estimate a multivariate multiple regression model. The significance level of these variables in the multivariate multiple regression model was checked again and the final model contained only those covariates that were statistically significant at least at the 0.1 level in the multivariable multiple regression model. All models were estimated in Stata (Version 13.0).

Results and Discussion

Descriptive Analysis

Table 4 shows that about 34 percent of the dwellings had no Internet connection in 2006 and that was reduced by 9 percent to 25 percent of the dwellings without an Internet connection in 2011. It also outlines that about one-fifth of the dwellings had access to dial-up Internet connection and that was also reduced to only 3 percent in 2011. In contrast, broadband Internet connection increased substantially from 23 percent in 2006 to 59 percent in 2011—an overall increase of 35 percent. Dwellings with other types of Internet connections, such as the Naked DSL, also increased by about 4 percent in that time. These changes suggest that there was an overall improvement with Internet connection in Australian dwellings from 2006 to 2011. A strong negative correlation (−0.79) between changes in broadband connection and changes in dial-up connection in Table 3 suggests that the LGAs that gave up dial-up connections gained accesses to broadband connections. Note that dwellings enjoy a better speed when connected by broadband and Naked DSL over the dial-up system.

Table 4. Patterns of change in Internet connection (2006–2011)

Type of Internet connection	2006 (% of dwellings)	2011 (% of dwellings)	Changes (%) between 2006 and 2011
Broadband connection	22.84	58.73	35.89
Dial-up connection	19.39	3.17	−16.22
Other types of Internet connection	0.51	4.26	3.75
No Internet connection	34.24	25.33	−8.91
Not applicable	16.60	1.46	−15.14
Not stated	6.42	7.05	0.63
Total (N = 513)	100	100	0.00

Table 5. Top 10 high-performing LGAs in terms of broadband Internet connection in Australia[a]

Top 10 LGAs in 2006		Top 10 LGAs in 2011		Top 10 LGAs with increasing connection	
LGA	% of dwellings	LGA	% of dwellings	LGA	% increase
Ku-ring-gai (A), NSW	62	Ku-ring-gai (A), NSW	84	Mallala (DC), SA	62
Willoughby (C), NSW	55	Nillumbik (S), VIC	82	Chittering (S), WA	62
Peppermint Grove (S), WA	55	Nedlands (C), WA	81	Woodanilling (S), WA	61
Hornsby (A), NSW	54	Hornsby (A), NSW	81	Wandering (S), WA	59
Lane Cove (A), NSW	52	Lane Cove (A), NSW	80	Wickepin (S), WA	57
Nillumbik (S), VIC	52	Willoughby (C), NSW	80	Dumbleyung (S), WA	57
Nedlands (C), WA	52	Pittwater (A), NSW	79	Kent (S), WA	56
Mosman (A), NSW	51	Joondalup (C), WA	79	Gingin (S), WA	56
Boroondara (C), VIC	50	Cambridge (T), WA	78	Lower Eyre Peninsula (DC), SA	56
Manningham (C), VIC	49	Boroondara (C), VIC	78	Narrogin (S), WA	56

[a]A = area, C = city, DC = district council, S = shire, T = town, NSW = New South Wales, SA = South Australia, VIC = Victoria, WA = Western Australia

Table 5 shows the top 10 LGAs in terms of broadband Internet connection both in 2006 and in 2011. Noticeable is that out of the 10 LGAs, seven remained in the top 10 list in both periods and also experienced a growth in broadband connection during that time. However, as expected, the maximum growth occurred in LGAs that were not in the top 10 list in 2006. The level of growth in these LGAs was found to be similar to the base level Internet connection of the top 10 LGAs. For example, Mallala (DC) in SA experienced a 62 percent increase in broadband connection between 2006 and 2011, whereas Ku-ring-gai LGA in NSW had 62 percent dwellings with broadband connection in 2006 (this was the highest level in 2006). Clearly this shows a distinction among the LGAs: (one having a legacy of high-level of broadband connection and the others that gained high-level connections between 2006 and 2011. This finding also justifies the inclusion of a "base" level variable in the multivariate multiple regression models (as discussed below) to investigate the effects of a legacy of Internet access on mode choice behavior.

Figure 1 shows the spatial distribution of LGAs in Australia with differential levels of change in broadband connection between 2006 and 2011. Clearly, the maximum level of changes occurred in LGAs located close to the capital cities—but not the capital city areas—perhaps because the capital city LGAs had already experienced a higher level of connection in 2006, and as a result, their level of change was minimal in that time. Further hot-spot analysis in ArcGIS, however, does not provide any spatial pattern of changes in broadband access in Australia.

Unlike changes in the exposure variable (access to the Internet), changes in outcome variables were found to be marginal. Table 6 shows that dependency on private transport increased by 2 percent between 2006 and 20ll; 61 percent of the employed people in Australia were dependent on a private vehicle for their travel to work in 2006 which was increased to 63 percent in 2011. In contrast, the rate of teleworking was reduced by 1.1 percent, from 9.6 percent in 2006 to 8.5 percent in 2011. Similarly, active transport use was reduced by 1.2 percent in that time. Eleven percent of the employed persons used

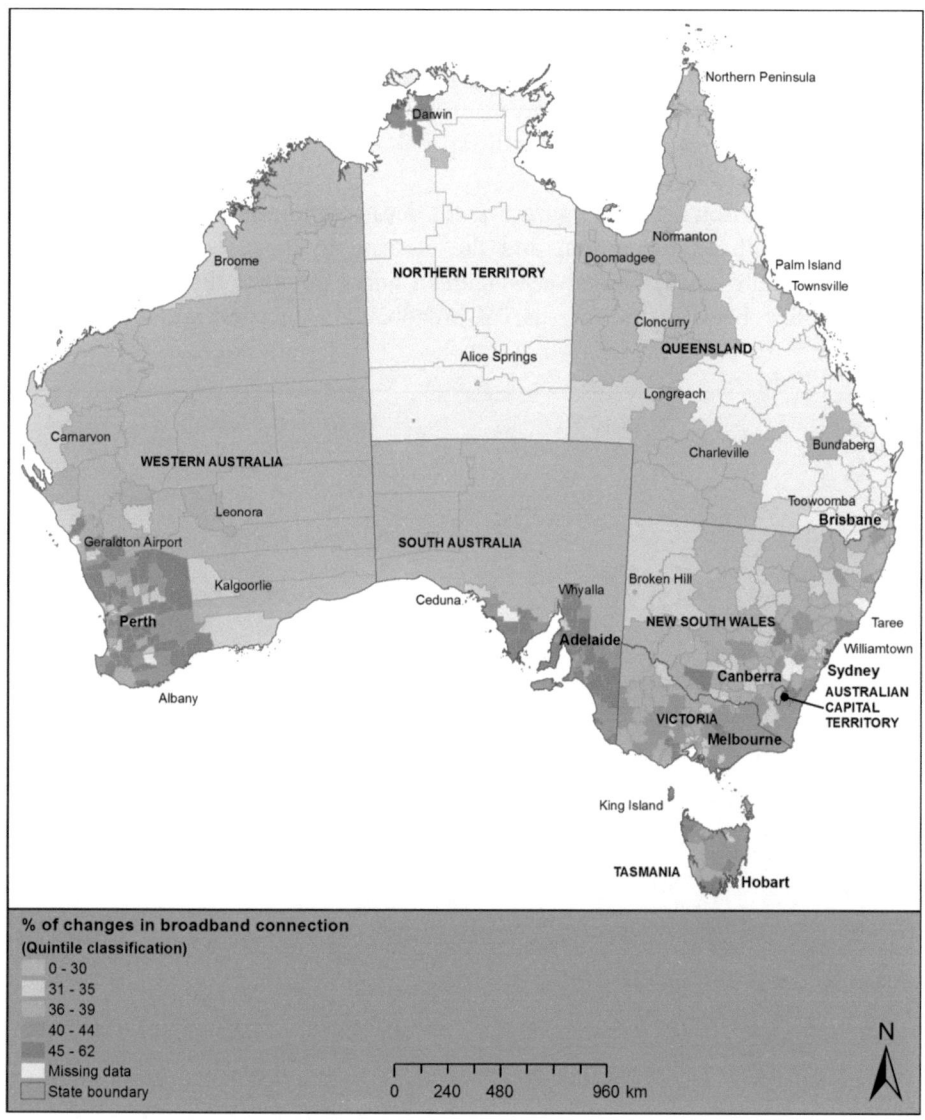

Figure 1. Changes of broadband Internet connection among the LGAs (2006–2011)

Table 6. Patterns of changes in commuting behavior (2006–2011)

	2006 (% of employed persons)	2011 (% of employed persons)	Changes (%) between 2006 and 2011
Worked at home	9.57	8.48	−1.10
Commuted by public transport	4.42	5.24	0.83
Commuted by private transport	60.99	63.01	2.02
Commuted by active transport	11.00	9.76	−1.23
Commuted by other mode	1.61	1.73	0.12
Did not go to work	10.34	10.02	−0.32
Not Stated	2.08	1.76	−0.31
Total	100	100	0.00

an active transport to work in 2006, which was reduced to 9.8 percent in 2011. All these findings point to a trend toward unsustainable commuting in Australia. The only sustainable growth pattern was evident in the use of public transport. The use of public transport for commuting increased by 0.83 percent from 4.42 percent in the base level to 5.24 percent in 2011. Figures 2 and 3 show the spatial distribution of changes in public and private transport uses, respectively.

A visual comparison between Figures 1 and 2 does not indicate a good link between the increase of broadband connection and the increase in the use of public transport. However, a visual comparison between Figures 1 and 3 points to the possibility of there being some link between the increase of broadband connection and the increase of

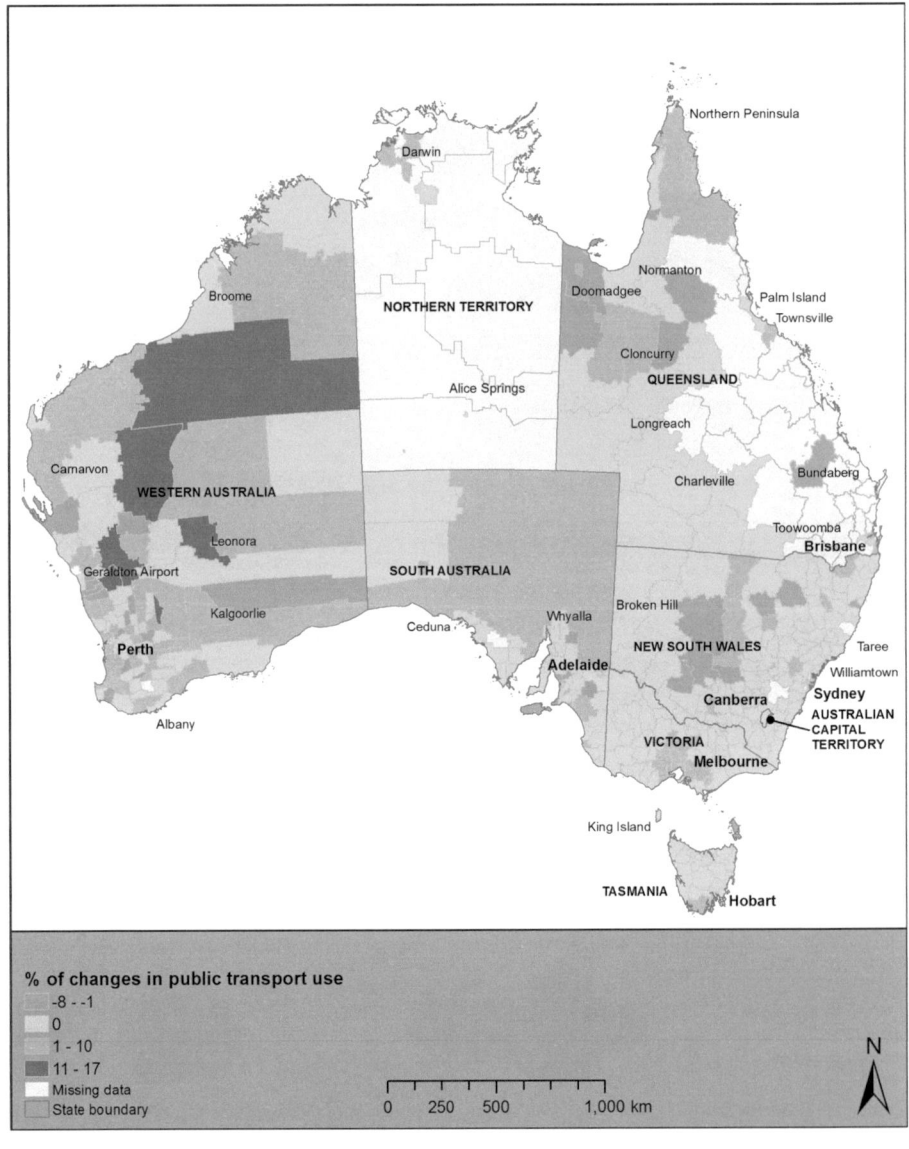

Figure 2. Changes in the use of public transport among the LGAs (2006–2011)

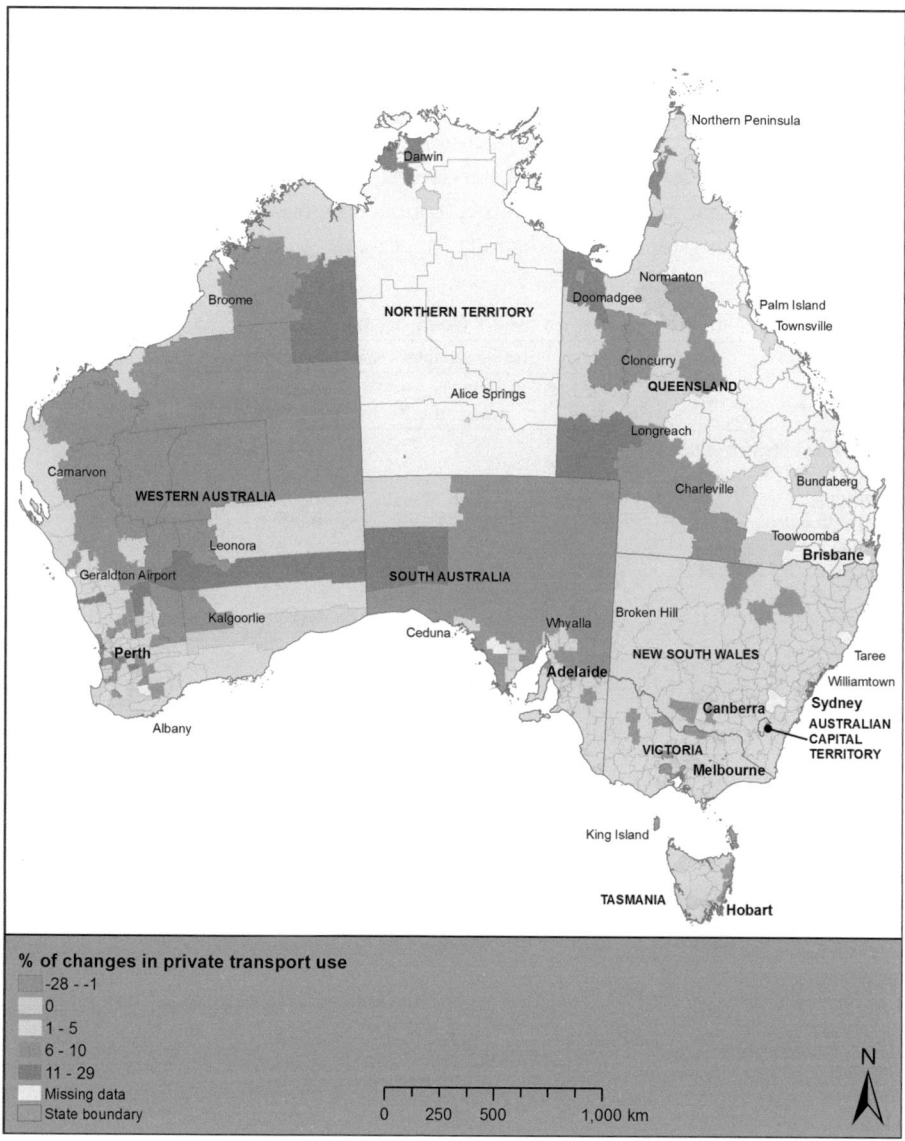

% of changes in private transport use
- -28 - -1
- 0
- 1 - 5
- 6 - 10
- 11 - 29
- Missing data
- State boundary

0   250   500   1,000 km

N

Figure 3. Changes in the use of private transport among the LGAs (2006–2011)

car-dependency. Attempts were made to ascertain these indicative relationships through multivariable multiple regression analysis as will be discussed next.

Multivariate Multiple Regression Analysis

This section reports the regression analysis results showing the factors that significantly affected the choice of commuter mode in Australia. This research conducted four different F tests (Wilks' lambda, Lawley-Hotelling trace, Pillai's trace, and Roy's largest root) to assess the statistical significance of the overall model. The test results for the overall model indicate that the multivariate models are statistically significant, regardless of the type of multivariate criteria used (e.g., Wilks' lambda). In addition, each of the five univariate models (e.g., the percentage of changes in worked at home, the percentage of changes in public transport use, and so on) was also found to be statistically significant (See Table 7). Moreover, the explanatory power (row labelled as R^2) of each of the univariate models was found to be quite favorable in comparison with previous research on this topic (Zhang et al., 2007). Note this value is a standard R-squared, not an adjusted R-squared.

Table 7 shows that broadband Internet connection has a significant effect on the choice of commuting mode in Australia. It reveals that a 1 percent increase in

Table 7. Multivariate multiple regression results (coefficients)

Explanatory factors	Outcome variables:[a] % of changes between 2006 and 2011 i				
	Worked at home	Public transport use	Private transport use	Active transport use	Other mode use
Change factors between 2006 and 2011					
% of changes i					
Dwellings with broadband connection	−0.02*	−0.03**	0.16**	−0.12**	−0.01*
Dwellings with other types of Internet connection	0.04	−0.21**	0.38**	−0.08	0.01
Dwellings with no Internet connection	0.04*	−0.07**	0.19**	−0.16**	0.01
Families with income of $1–$999	0.04**	−0.04**	0.01	0.04	0.00
Families with income of $2k–$2,999	−0.02	−0.02	0.04	−0.09	0.06**
Families with income of $3k–$3,999	0.04	−0.12**	−0.08	0.33**	0.00
Dwellings with zero vehicles	−0.09**	0.12**	−0.20**	0.38**	−0.11**
Dwellings with one vehicles	−0.04	0.22**	−0.11	0.04	0.04**
Dwellings with two vehicles	−0.10**	−0.03	0.55**	−0.24**	−0.02
Dwellings with three vehicles	0.20**	−0.04	−0.08	0.02	0.03
Dwellings with four or more vehicles	0.20**	0.12**	−0.18	−0.03	0.03
Female	0.29**	−0.66**	0.08	0.48**	−0.28**
Persons with full-time employment	−0.03*	0.04**	0.08*	0.01	0.02
Base factors in 2006					
% o					
Individuals worked at home	−0.23**	0.02	−0.02	0.25**	0.01
Individuals commuted by public transport	−0.02	0.06**	−0.12**	0.16**	0.01
Individuals commuted by private transport	−0.04**	0.03**	−0.15**	0.13**	0.00
Individuals did not go to work	0.04	0.13**	0.02	0.29**	0.03
Dwellings with other types of Internet connection	0.02	0.21**	0.16	−0.23	−0.29**
Dwellings with no Internet connection	−0.01	−0.02	0.15**	−0.06*	−0.03**
Families with negative or zero income	0.22**	0.28**	−0.21	−0.13	−0.03
Dwellings with one vehicles	0.05**	−0.07**	0.07	−0.07	0.04**
Dwellings with two vehicles	0.01	−0.04*	0.22**	−0.12**	0.01
Dwellings with four or more vehicles	0.09**	−0.07*	0.10	−0.16*	0.06**
Female	0.04	0.00	0.05	−0.08	−0.23**
Constant	−0.53	1.19	−9.57*	−0.25	11.52**
R^2	0.52	0.65	0.47	0.39	0.42
F	22.14**	37.30**	17.64**	12.92**	14.35**
N = 513 (local government areas)					

[a] Coefficients marked * are significant at the 0.1 level; Coefficients marked ** are significant at the 0.05 level

broadband connection reduced the number of people working from home by 0.02 percent. In contrast, an increasing number of households with no Internet connection increased the number of people working from home. A similar impact was found in the case of households with other types of Internet connection. The household Internet status in the base period (2006) was not found to be a statistically significant factor in the changes of commuting behavior in Australia. The findings, therefore, verify the replacement hypothesis of the ICT impact on travel behavior—but in an unsustainable way.

Even though the use of public transport increased overall between 2006 and 2011 as outlined earlier, Table 7 highlights that an increase in the level of broadband connection significantly reduced the use of public transport services. This is also true for households that increased their use of other types of Internet connections. Again, these findings relate to the replacement hypothesis of ICT impact on travel behavior—again in an unsustainable way. The complementary hypothesis is proved in the case of private transport use. Table 7 indicates that a 1 percent increase in the level of broadband connection in an LGA led to a 0.16 percent increase in private transport use. A similar impact is evident for other types of Internet connections. Access to the Internet may create fragmentation in work activities that need to be performed at different locations and at different times (Couclelis, 2000; Schwanen and Kwan, 2008), and naturally the most suitable mode to perform these fragmented activities is private transport, given their flexibility (Maat and Maat, 2009). Table 7 also highlights that a 1 percent increase in access to a broadband Internet connection led to a reduction in the use of active transport by 0.12 percent, which reflects the substitution hypothesis of ICT, but again in an unsustainable way.

Conclusion

The challenges our cities and societies have been facing in the age of global crises—environmental, economic, or social—have encouraged urban planners, architects, environmentalists, and policymakers to become passionate about new urban paradigms as potential panacea (Perveen et al., 2017; Yigitcanlar et al., 2017). As stated by Kunzmann (2014: 9),

> urban paradigms are urban dreamscapes, full of wishful thinking about better urban worlds. In the beginning of the twenty-first century, the sustainable city, the eco-city, the compact city, the creative city, the knowledge city, the slow city, the resilient city, and more recently, the smart city concept have received considerable academic interest, and attention among media and local governments, searching for popular visions for urban development in times of globalization.

Consequently, being "smart" is on the urban agenda of many cities across the globe with strong support from global technology and development companies—e.g., IBM, Cisco, Samsung, LG, ARUP, Schneider Electric, Siemens, Microsoft, Hitachi, Huawei, Ericsson, Toshiba, Oracle (Yigitcanlar, 2016; Alizadeh, 2017).

On the one hand, for many scholars, smart cities are seen as the immediate future, where smartness is perceived as a characteristic of city systems responding to opportunities, challenges, and unknown consequences (Albino et al., 2015). In contrast, sceptics argue that the smart cities movement should be considered with great caution as "large corporations are exerting significant influence in the era of smart in pursuit of goals that may not strongly align with those of urban planners concerned with social and environmental sustainability as well as economic prosperity" (Lyons, 2016: 1).

One of the key areas in which smartness is crucial for our cities is the transport sector. This importance has placed smart mobility at the heart of smart cities discourse and practice, where smart mobility is widely seen as connectivity in towns and cities that is affordable, effective, attractive, and sustainable (Garau et al., 2016). However, as argued by Lyons (2016), in mobility practice the paradigms of smart and of sustainable are not strongly aligned; the paradigms, thus, need to be brought together towards a common framework for truly smart and sustainable urban mobility development.

This paper investigated the relationship between urban smartness and sustainable forms of commuting in the context of Australian local government areas; in order to address the question of whether the smartness of cities leads to sustainable commuting patterns. The findings of this investigation generated several insights.

First, the descriptive analysis revealed, not to our surprise, that metropolitan regions of the capital cities have a higher-level of Internet (particularly broadband) accessibility and public and active transport use in Australia. Additionally, the level of working from home (or teleworking) has decreased between 2006 and 2011. Despite metropolitan regions having higher public and active transport shares, they are still too low, and the figures point to a clear unsustainable commuting pattern. The visible correlation between Internet access and commuting is as follows: an increase in broadband access leads to an increase in private motor vehicle use.

Second, the multivariate multiple regression analysis revealed that the Internet (particularly broadband) has a significant effect on commuting. An increase in broadband access in a locality decreased teleworking—even though marginally—and affected travel behavior in an unsustainable way. For instance, an increase in broadband has negative effects both on public transport and active transport uses and increases private motor vehicle use. This finding underlines the negative relationship between Internet access (or smartness in this study) and sustainable commuting patterns. In sum, the overall results highlight that an increasing access to the broadband Internet reduces the level of working from home, public transport use, and active transport use; but at the same time increases the use of private vehicles perhaps to overcome the fragmentation of work activities.

Third, even though sustainable urban development and sustainable transport concepts have been widely investigated (e.g., Kamruzzaman et al., 2015; Yigitcanlar and Dizdaroglu, 2015; Yigitcanlar and Teriman, 2015), smart cities and smart mobility concepts have not been extensively empirically investigated in Australia and overseas. Therefore, further empirical investigations are needed to clearly address the issue of whether the smartness of cities leads to sustainable commuting patterns—in other words, the relationship between "smart" and "sustainable." This issue has recently been investigated by Yigitcanlar and Kamruzzaman (2018) in the context of cities in the United Kingdom. They found no clear evidence that smart city policies lead to sustainable cities. In the light of the aforementioned findings, a key to the smart cities agenda in Australia should be the creation of strategies needed to overcome the need for car-based travel for fragmented work activities, while increasing smartness through the provisioning of broadband access. This is a critical issue in order to make Australian cities more sustainable.

Fourth, the aforementioned findings should be considered carefully as the study has limitations. These limitations include: (a) due to data availability and boundary change issues, the study has not used the latest Census data—i.e., 2016. The study might not be able to capture the recent trends as in the age of digital disruption, technologies—e.g., social media—are rapidly altering the behaviors of individuals and societies; (b) using Internet access as the sole indicator of urban smartness has its limitations. As smart cities, urban smartness is also a fuzzy concept and using composite indicators to determine the smartness might be a better approach (Yigitcanlar et al., 2018); (c) the findings would benefit from ground-truthing by discussing them with experts and a group of officials from the investigated local government areas; (d) the data used in this research is panel in nature, but not a true panel data, and therefore, it was not possible to investigate the impacts at the individual level. Consequently, the estimated coefficients might capture effects that are not properly disentangled in this research. For example, previously "not stated" individuals may become part of the "broadband connection" group although they might have had a broadband connection previously.

Finally, considering one of the findings of this study—as individual levels of broadband access increase, commuting preferences lean towards private motor vehicle use—smart mobility technologies may have something to offer. Investments in autonomous vehicle technology may in the long run provide more sustainable, and at the same time, more convenient and comfortable commuting options (Firnkorn and Müller, 2015; Greenblatt and Saxena, 2015). Although it is unlikely to happen anytime soon, shared or pooled electric autonomous vehicles might decrease car ownership and enhance active transport use—as the freed space from cars could be redesigned for a whole new spectrum of social functions, street trees, walkways, or bike lanes (Wadud et al., 2016; Meyer et al., 2017).

Acknowledgments

The authors are grateful to the editor and anonymous reviewers, who provided constructive comments on an earlier version of the paper.

ORCID

Tan Yigitcanlar ⓘ http://orcid.org/0000-0001-7262-7118
Md. Kamruzzaman ⓘ http://orcid.org/0000-0001-7113-942X

Bibliography

A. Aguiléra, C. Guillot, and A. Rallet, "Mobile ICTs and Physical Mobility: Review and Research Agenda," *Transportation Research Part A* 46 (2012) 664–672.

V. Albino, U. Berardi, and R. Dangelico, "Smart Cities: Definitions, Dimensions, Performance, and Initiatives," *Journal of Urban Technology* 22 (2015) 3–21.

B. Alexander, D. Ettema, and M. Dijst, "Fragmentation of Work Activity as a Multi-Dimensional Construct and its Association with ICT, Employment and Sociodemographic Characteristics," *Journal of Transport Geography* 18 (2010) 55–64.

T. Alizadeh, "An Investigation of IBM's Smarter Cites Challenge: What Do Participating Cities Want?" *Cities* 63 (2017) 70–80.

P. Allison, *Fixed Effects Regression Models* (Thousand Oaks CA: Sage, 2009).

W. Anderson, L. Chatterjee, and T. Lakshmanan, "E-commerce, Transportation, and Economic Geography," *Growth and Change* 34 (2003) 415–432.

M. Angelidou, "Smart City Policies: a Spatial Approach," *Cities* 41 (2014) S3–S11.

M. Angelidou, "The Role of Smart City Characteristics in the Plans of Fifteen Cities," *Journal of Urban Technology* 24 (2017) 3–28.

L. Anthopoulos, "Smart Utopia vs Smart Reality: Learning by Experience from 10 Smart City Cases," *Cities* 63 (2017) 128–148.

A. Antipova, F. Wang, and C. Wilmot, "Urban Land Uses, Socio-Demographic Attributes and Commuting: a Multilevel Modeling Approach," *Applied Geography* 31 (2011) 1010–1018.

R. Arbolino, F. Carlucci, A. Cirà, G. Ioppolo, and T. Yigitcanlar, "Efficiency of the EU Regulation on Greenhouse Gas Emissions in Italy: The Hierarchical Cluster Analysis Approach," *Ecological Indicators* 81 (2017) 115–123.

J. Asensio, "Transport Mode Choice by Commuters to Barcelona's CBD," *Urban Studies* 39 (2002) 1881–1895.

Australian Government, *Smart Cities Plan* (Canberra: Department of the Prime Minister and Cabinet, 2016).

M. Batty, "Big Data, Smart Cities and City Planning," *Dialogues in Human Geography* 3 (2013) 274–279.

M. Batty, K. Axhausen, F. Giannotti, A. Pozdnoukhov, A. Bazzani, M. Wachowicz, and Y. Portugali, "Smart Cities of the Future," *The European Physical Journal Special Topics* 214 (2012) 481–518.

S. Baum, Y. Van Gellecum, and T. Yigitcanlar, "Wired Communities in the City: Sydney, Australia," *Geographical Research* 42 (2004) 175–192.

J. Beckman and K. Goulias, "Immigration, Residential Location, Car Ownership, and Commuting Behavior: a Multivariate Latent Class Analysis from California," *Transportation* 35 (2008) 655–671.

M. Bowles and P. Wilson, "The NBN and Australia's Race to Compete in the Digital Economy," *Australian Quarterly* 83 (2012) 11–19.

C. Brown, P. Balepur, and P. Mokhtarian, "Communication Chains: A Methodology for Assessing the Effects of the Internet on Communication and Travel," *Journal of Urban Technology* 12 (2005) 71–98.

Z. Bursac, C. Gauss, D. Williams, and D. Hosmer, "Purposeful Selection of Variables in Logistic Regression," *Source Code for Biology and Medicine* 3:7 (2008).

X. Cao, "Heterogeneous Effects of Neighborhood Type on Commute Mode Choice: An Exploration of Residential Dissonance in the Twin Cities," *Journal of Transport Geography* 48 (2015) 188–196.

D. Chang, J. Sabatini-Marques, E. da Costa, P. Selig, and T. Yigitcanlar, "Knowledge-based, Smart and Sustainable Cities: A Provocation for a Conceptual Framework," *Journal of Open Innovation: Technology, Market, and Complexity* 4:5 (2018).

B. Chun and S. Lee, "Review on ITS in Smart City," *Advanced Science and Technology Letters* 98 (2015) 52–54.

G. Claisse and F. Rowe, "Domestic Telephone Habits and Daily Mobility," *Transportation Research Part A* 27 (1993) 277–290.

B. Clark, K. Chatterjee and S. Melia, "Changes to Commute Mode: the Role of Life Events, Spatial Context and Environmental Attitude," *Transportation Research Part A* 89 (2016) 89–105.

G. Cohen-Blankshtain, and O. Rotem-Mindali, "Key Research Themes on ICT and Sustainable Urban Mobility," *International Journal of Sustainable Transportation* 10 (2016) 9–17.

H. Couclelis, "From Sustainable Transportation to Sustainable Accessibility: Can We Avoid a New Tragedy of the Commons?" in D. Janelle and D. Hodge, eds, *Information, Place, and Cyberspace* (Berlin: Springer, 2000) 341–356.

F. Creutzig, N. Ravindranath, G. Berndes, S. Bolwig, R. Bright, F. Cherubini, H. Chum, E. Corbera, M. Delucchi, A. Faaij, and J. Fargione, "Bioenergy and Climate Change Mitigation: An Assessment." *Gcb Bioenergy* 7 (2015) 916–944.

P. Dattalo, *Analysis of Multiple Dependent Variables* (New York: Oxford University Press, 2013).

O. Fadare and B. Salami, "Telephone Uses and the Travel Behaviour of Residents in Osogbo, Nigeria: an Empirical Analysis," *Journal of Transport Geography* 12 (2004) 159–164.

J. Firnkorn and M. Müller, "Free-floating Electric Carsharing-Fleets in Smart Cities: The Dawning of a Post-Private Car Era in Urban Environments?" *Environmental Science and Policy* 45 (2015) 30–40.

C. Garau, F. Masala, and F. Pinna, "Cagliari and Smart Urban Mobility: Analysis and Comparison," *Cities* 56 (2016) 35–46.

D. Glance, "When It Comes to the NBN, We Keep Having the Same Conversations Over and Over" (Melbourne: The Conservation, 2017) <https://theconversation.com> Accessed October 6, 2017.

P. Goldmark, *Telecommunications for Enhanced Metropolitan Function and Form* (Washington DC: National Academy of Engineering, 1969).

A. Goonetilleke, T. Yigitcanlar, G. Ayoko, and P. Egodawatta, *Sustainable Urban Water Environment: Climate, Pollution and Adaptation* (Cheltenham: Edward Elgar, 2014).

J. Greenblatt and S. Saxena, "Autonomous Taxis Could Greatly Reduce Greenhouse-Gas Emissions of US Light-Duty Vehicles," *Nature Climate Change* 5 (2015) 860–863.

S. Handy, X. Cao, and P. Mokhtarian, "Correlation or Causality Between the Built Environment and Travel Behavior? Evidence from Northern California," *Transportation Research Part D* 10 (2005) 427–444.

S. Handy and T. Yantis, *The Impacts of Telecommunications Technologies on Non-Work Travel Behavior* (Texas: Southwest Region University, 1997).

R. Hjorthol, "The Relation between Daily Travel and Use of the Home Computer," *Transportation Research Part A* 36 (2002) 437–452.

T. Hortz, "The Smart State Test: A Critical Review of the Smart State Strategy 2005–2015's Knowledge-Based Urban Development," *International Journal of Knowledge-Based Development* 7 (2016) 75–101.

D. Hosmer, S. Lemeshow, and R. Sturdivant, *Applied Logistic Regression* (New Jersey: Wiley, 2013).

A. Johansson, "Transport in an Era of Communication," *SIKA Document* 1 (1999) 1–10.

M. Kamruzzaman and J. Hine, "Self-proxy Agreement and Weekly School Travel Behaviour in a Sectarian Divided Society," *Journal of Transport Geography* 29 (2013) 74–85.

M. Kamruzzaman, J. Hine, F. Shatu, and G. Turrell, "Commuting Mode Choice Behaviour in TODS, TADS, and Traditional Suburbs," paper presented at the 46th Annual Universities' Transport Study Group Conference (Newcastle, October 6–9, 2014).

M. Kamruzzaman, J. Hine, and T. Yigitcanlar, "Investigating the Link Between Carbon Dioxide Emissions and Transport Related Social Exclusion in Rural Northern Ireland," *International Journal of Environmental Science and Technology* 12 (2015) 3463–3478.

M. Kamruzzaman, S. Washington, D. Baker, W. Brown, B. Giles-Corti, and G. Turrell, "Built Environment Impacts on Walking for Transport in Brisbane, Australia," *Transportation* 43 (2016) 53–77.

S. Kenyon "The Impacts of Internet Use Upon Activity Participation and Travel: Results from a Longitudinal Diary-Based Panel Study," *Transportation Research Part C* 18 (2010) 21–35.

J. Kim, Y. Moon, and I. Suh, "Smart Mobility Strategy in Korea on Sustainability, Safety and Efficiency Toward 2025," *IEEE Intelligent Transportation Systems Magazine* 7 (2015) 58–67.

N. Komninos, "Smart Environments and Smart Growth: Connecting Innovation Strategies and Digital Growth Strategies," *International Journal of Knowledge-Based Development* 7 (2016) 240–263.

K. Krizek, "Residential Relocation and Changes in Urban Travel: Does Neighborhood-Scale Urban Form Matter?" *Journal of the American Planning Association* 69 (2003) 265–281.

K. Kunzmann, "Smart Cities: A New Paradigm of Urban Development," *Crios* 1 (2014) 9–20.

M. Kwan, M. Dijst, and T. Schwanen, "The Interaction between ICT and Human Activity-Travel Behavior," *Transportation Research Part A* 41 (2007) 121–124.

A. Lara, E. Costa, T. Furlani, and T. Yigitcanlar, "Smartness that Matters: Towards a Comprehensive and Human-Centred Characterisation of Smart Cities," *Journal of Open Innovation: Technology, Market, and Complexity* 2:8 (2016).

J. Lee, M. Hancock, and M. Hu, "Towards an Effective Framework for Building Smart Cities: Lessons from Seoul and San Francisco," *Technological Forecasting & Social Change* 89 (2014) 80–99.

P. Lila and M. Anjaneyulu, "Modeling the Impact of ICT on the Activity and Travel Behaviour of Urban Dwellers in Indian Context," *Transportation Research Procedia* 17 (2016) 418–427.

T. Line, J. Jain, and G. Lyons, "The Role of ICTs in Everyday Mobile Lives," *Journal of Transport Geography* 19 (2011) 1490–1499.

G. Lyons, "Internet: Investigating New Technology's Evolving Role, Nature and Effects on Transport," *Transport Policy* 9 (2002) 335–346.

G. Lyons, "Getting Smart About Urban Mobility: Aligning the Paradigms of Smart and Sustainable," Transportation Research Part A (2016) <https://doi.org/10.1016/j.tra.2016.12.001>

C. Maat and K. Maat, *Built Environment and Car Travel: Analyses of Interdependencies* (New York: IOS Press, 2009).

P. Mahbub, A. Goonetilleke, G. Ayoko, P. Egodawatta, and T. Yigitcanlar, "Analysis of Build-Up of Heavy Metals and Volatile Organics on Urban Roads in Gold Coast, Australia," *Water Science and Technology* 63 (2011) 2077–2085.

J. Messenger and L. Gschwind, "Three Generations of Telework: New ICTs and the (R)Evolution from Home Office to Virtual Office," *New Technology, Work and Employment* 31 (2016) 195–208.

J. Meyer, H. Becker, P. Bösch, and K. Axhausen, "Autonomous Vehicles: The Next Jump in Accessibilities?" *Research in Transportation Economics* 62 (2017) 80–91.

P. Mokhtarian, "Telecommuting and Travel: State of the Practice, State of the Art," *Transportation* 18 (1991) 319–342.

P. Mokhtarian, "A Conceptual Analysis of the Transportation Impacts of B2C E-commerce," *Transportation* 31 (2004) 257–284.

P. Neirotti, A. De Marco, A. Cagliano, G. Mangano, and F. Scorrano, "Current Trends in Smart City Initiatives: Some Stylised Facts," *Cities* 38 (2014) 25–36.

C. Nobis and B. Lenz, "Changes in Transport Behavior by Fragmentation of Activities," *Transportation Research Record* 1894 (2004) 249–257.

OECD, *Access Network Speed Tests* (Paris: OECD Publishing, 2014).

S. Perveen, M. Kamruzzaman, and T. Yigitcanlar, "Developing Policy Scenarios for Sustainable Urban Growth Management: A Delphi Approach," *Sustainability* 9:1787 (2017).

L. Piya, K. Maharjan, and N. Joshi, "Determinants of Adaptation Practices to Climate Change by Chepang Households in the Rural Mid-Hills of Nepal," *Regional Environmental Change* 13 (2013) 437–447.

S. Praharaj, J. Han, and S. Hawken, "Urban Innovation Through Policy Integration: Critical Perspectives from 100 Smart Cities Mission in India," *City, Culture and Society* 12 (2018) 35–43.

F. Ren and M. Kwan, "The Impact of the Internet on Human Activity-Travel Patterns: Analysis of Gender Differences Using Multi-Group Structural Equation Models," *Journal of Transport Geography* 17 (2009) 440–450.

O. Rotem-Mindali, "E-tail Versus Retail: The Effects on Shopping Related Travel Empirical Evidence from Israel," *Transport Policy* 17 (2010) 312–322.

T. Schwanen and M. Kwan, "The Internet, Mobile Phone and Space-Time Constraints," *Geoforum* 39 (2008) 1362–1377.

J. Spinney, D. Scott, and K. Newbold, "Transport Mobility Benefits and Quality of Life: a Time-Use Perspective of Elderly Canadians," *Transport Policy* 16 (2009) 1–11.

K. Srinivasan and A. Reddy, "Modeling Interaction Between Internet Communication and Travel Activities: Evidence from Bay Area, California, Travel Survey 2000," *Transportation Research Record* 1894 (2004) 230–240.

A. Taamallah, M. Khemaja, and S Faiz, "Strategy Ontology Construction and Learning: Insights from Smart City Strategies," *International Journal of Knowledge-Based Development* 8 (2017) 206–228.

A. Townsend, *Smart Cities: Big Data, Civic Hackers, and the Quest for a New Utopia* (New York: WW Norton, 2013).

E. Trindade, M. Hinnig, E. Costa, J. Sabatini-Marques, R. Bastos, and T. Yigitcanlar, "Sustainable Development of Smart Cities: A Systematic Review of the Literature," *Journal of Open Innovation: Technology, Market, and Complexity* 3:11 (2017).

L. Truong, C. De Gruyter, G. Currie, and A. Delbosc, "Estimating the Trip Generation Impacts of Autonomous Vehicles on Car Travel in Victoria, Australia," *Transportation* 44 (2017) 1279–1292.

R. Tucker and P. Branch, "Fact Check: Will the NBN Take Another 20 Years to Complete?" (Melbourne: The Conservation, 2013) <https://theconversation.com> Accessed October 6, 2017.

J. Twisk, *Applied Longitudinal Data Analysis for Epidemiology: A Practical Guide* (Cambridge: Cambridge University Press, 2003).

K. Viswanathan and K. Goulias, "Travel Behavior Implications of Information and Communications Technology in Puget Sound Region," *Transportation Research Record* 1752 (2001) 157–165.

Z. Wadud, D. MacKenzie, and P. Leiby, "Help or Hindrance? The Travel, Energy and Carbon Impacts of Highly Automated Vehicles," *Transportation Research Part A* 86 (2016) 1–18.

A. Wiig, "IBM's Smart City as Techno-Utopian Policy Mobility," *City* 19 (2015) 258–273.

W. Wilson, *P. Sa, and R. Freund, Regression Analysis: Statistical Modeling of a Response Variable* (London: Academic Press, 2006).

T. Yigitcanlar, "Is Australia Ready to Move Planning to Online Mode?" *Australian Planner* 42 (2005) 42–51.

T. Yigitcanlar, "Australian Local Governments' Practice and Prospects with Online Planning," *URISA Journal* 18 (2006) 7–17.

T. Yigitcanlar, "Smart Cities: An Effective Urban Development and Management Model?" *Australian Planner* 52 (2015) 27–34.

T. Yigitcanlar, *Technology and the City: Systems, Applications and Implications* (New York: Routledge, 2016).

T. Yigitcanlar, "Smart Cities in the Making," *International Journal of Knowledge-Based Development* 8 (2017) 201–205.

T. Yigitcanlar and D. Dizdaroglu, "Ecological Approaches in Planning for Sustainable Cities: A Review of the Literature," *Global Journal of Environmental Science and Management* 1 (2015) 71–94.

T. Yigitcanlar, D. Dur, and D. Dizdaroglu, "Towards Prosperous Sustainable Cities: A Multiscalar Urban Sustainability Assessment Approach," *Habitat International* 45 (2015) 36–46.

T. Yigitcanlar, I. Edvardsson, H. Johannesson, M. Kamruzzaman, G. Ioppolo, and S. Pancholi, "Knowledge-Based Development Dynamics in Less Favoured Regions: Insights from Australian and Icelandic University Towns," *European Planning Studies* 25 (2017) 2272–2292.

T. Yigitcanlar and M. Kamruzzaman, "Investigating the Interplay between Transport, Land Use and the Environment: A Review of the Literature," *International Journal of Environmental Science and Technology* 11 (2014) 2121–2132.

T. Yigitcanlar and M. Kamruzzaman, "Planning, Development and Management of Sustainable Cities: A Commentary from the Guest Editors," *Sustainability* 7 (2015) 14677–14688.

T. Yigitcanlar and M. Kamruzzaman, "Does Smart City Policy Lead to Sustainability of Cities?" *Land Use Policy* 73 (2018) 49–58.

T. Yigitcanlar, M. Kamruzzaman, L. Buys, G. Ioppolo, J. Sabatini-Marques, E. Costa and J. Yun, "Understanding Smart Cities: Intertwining Development Drivers with Desired Outcomes in a Multidimensional Framework," *Cities* (2018) https://doi.org/10.1016/j.cities.2018.04.003.

T. Yigitcanlar and S. Lee, "Korean Ubiquitous-Eco-City: A Smart-Sustainable Urban Form or a Branding Hoax?" *Technological Forecasting and Social Change* 89 (2014) 100–114.

T. Yigitcanlar and S. Teriman, "Rethinking Sustainable Urban Development: Towards an Integrated Planning and Development Process," *International Journal of Environmental Science and Technology* 12 (2015) 341–352.

F. Zhang, K. Clifton, and Q Shen, "Reexamining ICT Impact on Travel Using the 2001 NHTS Data for Baltimore Metropolitan Area," in H. Miller, ed., *Societies and Cities in the Age of Instant Access* (Berlin: Springer, 2007) 153–166.

The (In)Security of Smart Cities: Vulnerabilities, Risks, Mitigation, and Prevention

Rob Kitchin [ID] and Martin Dodge

ABSTRACT
In this paper we examine the current state of play with regards to the security of smart city initiatives. Smart city technologies are promoted as an effective way to counter and manage uncertainty and urban risks through the effective and efficient delivery of services, yet paradoxically they create new vulnerabilities and threats, including making city infrastructure and services insecure, brittle, and open to extended forms of criminal activity. This paradox has largely been ignored or underestimated by commercial and governmental interests or tackled through a technically-mediated mitigation approach. We identify five forms of vulnerabilities with respect to smart city technologies, detail the present extent of cyberattacks on networked infrastructure and services, and present a number of illustrative examples. We then adopt a normative approach to explore existing mitigation strategies, suggesting a wider set of systemic interventions (including security-by-design, remedial security patching and replacement, formation of core security and computer emergency response teams, a change in procurement procedures, and continuing professional development). We discuss how this approach might be enacted and enforced through market-led and regulation/management measures, and then examine a more radical preventative approach to security.

Introduction

Over the past two decades there has been a concerted move to network urban infrastructures to utilize computation to try to solve urban problems and deliver city services more efficiently. Such endeavors are now encapsulated within the notion of smart cities, a world-wide movement that seeks to transform urban governance, management, and living through the use of new networked digital technologies. For advocates, the creation of smart cities will help address issues of urban resilience and sustainability in a time of rapid population increases, environmental change, and fiscal austerity (see Söderström et al., 2014; White, 2016). In other words, smart city technologies are seen to offer an effective way to counter and manage uncertainty and risk. However, as with previous rounds of technological adoption and adaptation in cities (such as those related to energy supply, transportation systems, communication services), they also create a paradoxical situation wherein the promised benefits (such as convenience, economic prosperity, safety, sustainability) are accompanied by unintended consequences and new variances of traditional problems (e.g., reproducing inequality, creating security and criminal risks, environmental externalities) (See Datta, 2015; Greenfield, 2013; Singh and Pelton, 2013; Townsend, 2013). This paradoxical relationship and the reproduction of urban problems and risks in a new guise is for the most part ignored in the promotional discourse for smart cities driven by commercial and governmental interests or is present as a potential new issue to be "solved" by a further round of technological innovation and capital spending.

In contrast, in this paper we examine this paradoxical relationship in depth, detailing how smart city technologies designed to produce urban resilience and reduce risks are actually opening up the urban systems they are meant to augment to new forms of vulnerability and risk. In particular, we are interested in considering the balancing point between reward and risk when previously relatively "dumb" systems are made "smart" through the introduction of networked computation, and are thus opened up to software bugs, data errors, network viruses, hacks, and criminal and terrorist enterprise (Little, 2010; Kitchin and Dodge, 2011; Townsend, 2013; Cerrudo, 2015). We are especially interested in security vulnerabilities and the extent to which it is becoming possible to hack and disrupt smart city technologies and to commit new variances of criminal activity.

As the burgeoning literature on crime and the city details, for as long as there have been urban societies there has been criminal activity and attempts to penetrate, attack, defraud, and disrupt city infrastructure and public services (Evans and Herbert, 1989; LeBeau and Leitner, 2011; Hall, 2013). Attempts to limit and defend against such crimes have become built into the fabric of cities themselves through architecturally enacted defenses, strong doors, locks, window grills, high walls and fences, security alarms, and CCTV (Manaugh, 2016). However, history has shown that all these security measures have some vulnerabilities that criminals are quick to identify and exploit. With time, all security, even sophisticated or well-designed solutions, will be defeated (especially if the reward of success provides sufficient motivation). There is thus a perennial struggle between defenders and attackers to secure systems that provide adequate protection but are not so restrictive that they seriously inconvenience users or inhibit essential economic transactions.

Smart city technologies are no different, being afflicted with a range of security vulnerabilities and risks, and an ongoing struggle is now evident between the cybersecurity industry and criminals and variously motivated hackers. However, while the base motivations to break into these systems might remain timeless (e.g., theft, extortion, impersonation, vandalism, malicious attack: See Schneier, 2003), the nature of their performance is different. Because smart city technologies rely on networked digital computation, exploits of their vulnerabilities can be undertaken at distance and attacks can be masked, reducing the risk of detection and capture for perpetrators. Moreover, the use of software tools to automate hacking has greatly lowered-costs and "super-empowered" individual actors to conduct virtual criminality against multiple targets simultaneously, potentially affecting many different cities. Unauthorized access is often made easier because the so-called "attack surfaces"— the set of ways that a system might be susceptible to an attack (Bellovin, 2016)—are multiplied due to a system's many interlocking parts, which are owned and controlled by a diverse set of stakeholders, making it difficult to secure every aspect of a large infrastructure or utility network (Article 29 DPWP, 2014; Cerrudo, 2015; Durbin, 2015). The rewards for success can also be significant, for example in the case of a data breach providing access to millions of user details, or in the case of vandalism/terrorism shutting down the entire electricity supply to a city, and can garner large amounts of publicity.

In the first part of the paper we detail the various security vulnerabilities of smart cities and their associated risks, providing contemporary illustrative examples from European and North American cities. In the second part, we chart the ways in which these vulnerabilities and risks are being tackled through mitigation strategies, how these strategies might be further encouraged and complemented by market-based and governance-based incentives and regulation, and consider a more radical preventative strategy. Our approach is *normative*. Rather than providing another critique of the smart city, this time by charting the paradoxical situation in which technologies designed to tackle urban problems are introducing new vulnerabilities and risks, or seeking to frame such risks within the discourses of the risk society (Beck, 1992), or urban resilience (Mackinnon and Derickson, 2013), or the political economy of neoliberal smart urbanism (Kitchin, 2014; Luque-Ayala and Marvin, 2016), we are more interested in examining the security challenges and threats faced by cities today and in the coming decade and how they might be more effectively mitigated and prevented. Our approach is guided by a recognition that the use of software systems and networked technologies to manage urban services and govern cities is firmly established and is only likely to become more entrenched, and by the need for a coherent approach to security that extends beyond technical solutions.

Forms of Cyberattack and Amplifying Factors

Cyberattacks seek to "alter, disrupt, deceive, degrade, or destroy computer systems and networks or the information and/or programs resident in or transiting these systems or networks" (Owens et al., 2009: 1). There are three distinct forms of cyberattack against operational systems: *availability attacks* that seek to close a system down or deny service use; *confidentiality attacks* that seek to extract information and monitor activity; and *integrity attacks* that seek to enter a system to alter information and settings (such as changing settings so that components exceed normal performance, erasing critical software, planting malware and viruses) without being noticed by the legitimate operator/owner (Singer and Friedman, 2014). Cyberattacks can be performed by multiple different actors, from nation state intelligence agencies and militaries, terrorist groups, organized criminals, hacker collectives, political and socially motivated activists to "lone wolf" hackers, "script kiddies" and bored teenagers.[1] Former FBI Director, Robert Mueller, has claimed that 108 nations have government funded and directed cyberattack units, targeting critical infrastructure and industrial secrets (Goodman, 2015). Anecdotal evidence from media reporting indicates a significant increase in organized criminals conducting thefts and frauds by targeting online systems, including spate of so-called "randomware" attacks against organizations (Hern, 2016).

In general, cyberattacks try to exploit one of five major vulnerabilities of digital technologies that are central to smart city systems. The first of these is *weak software security and data encryption*. Research by a Carnegie Mellon University team in 2004 detailed that, on average, there are 30 errors or possibly exploitable bugs for every 1,000 lines of code (Li et al., 2004). In typical large systems being deployed in cities there are millions of lines of code that produce thousands of potential zero-day exploits (as yet unknown security vulnerabilities) for network viruses, malware, and directed hacks. Moreover, research by cybersecurity specialists has detailed how many smart city systems have been constructed with no or minimal security (Cerrudo, 2015). For example, using the *Shodan* search engine (www.shodan.io: See Bodenheim et al., 2014) it is possible to find all kinds of digital devices and control systems connected to the Internet—from networked thermostats for heating systems to traffic control systems and command-and-control centers for nuclear power plants—many of which have been found to have little to no security (such as no user authentication, or using default or weak passwords, e.g., "admin," "1234"). Moreover, city governments and vendors of smart city technologies often deploy them without undertaking cybersecurity testing (Cerrudo, 2015). In the case of some "Internet of Things" (IoT)[2] deployments, it can be difficult to ensure end-to-end security because most sensors and low-powered devices on the market do not have sufficient computing power to support an encrypted network link (Article 29 DPWP, 2014). Where encryption is used, security issues can arise regarding how it is operated (Cerrudo, 2015).

The second area of vulnerability concerns the *use of insecure legacy systems and poor ongoing maintenance*. Many smart city technologies are layered onto much older infrastructure that relies on software and technology created 20 or 30 years ago, which has not been upgraded for some time, nor can they be migrated to newer, more secure systems (Rainie et al., 2014; Cerrudo, 2015). These technologies can create inherent vulnerabilities to newer systems by providing so-called "forever-day exploits" (holes in legacy software products that vendors no longer support and thus will never be patched) (Townsend, 2013). Even in the case of newer technologies, it can be difficult to test and rollout patches onto critical operational systems that need to always be on (Cerrudo, 2015).

The third vulnerability is that smart city systems are typically large, complex and diverse, with *many interdependencies and large and complex attack surfaces*. Such complexity means it can be difficult to know what and how all the components are exposed, to measure and mitigate risks, and to ensure end-to-end security (Article 29 DPWP, 2014; Cerrudo, 2015; Durbin, 2015). Even if independent systems are secure, linking them to other systems can potentially open them to risk with the level of security only guaranteed by the weakest link. Moreover, the interdependencies between technologies and systems mean that they are harder to maintain and upgrade (Sarma, 2015). Beyond being hacked, the complexity of systems also increases the chances of "normal accidents" (e.g., programming bugs, human errors) that cause unanticipated failures (Perrow, 1984; Townsend, 2013).

The interdependencies between smart city technologies and systems have the potential to create *cascade effects*, wherein "highly interconnected entities rapidly transmit adverse consequences to each other" (Durbin, 2015; see also Little, 2010). For example, a cyberattack on an electrical power infrastructure could cascade into an urban operating system that then cascades into the other systems such as traffic management, emergency services, and water services. Indeed, this is one of the key security and resilience risks of an urban operating system, wherein several systems are linked together to enable a "system of systems" approach to managing city services and infrastructures thus undoing the mitigating effects of using a siloed approach (i.e., fully separate system with physically independent cabling and sources of power, etc.) (Little, 2010). A successful cyberattack on the electricity grid has huge cascade effects as it underpins so many activities such as powering homes, workplaces, and a plethora of other essential infrastructures. For example, a sophisticated cyberattack on the software controlling parts of Ukraine's electricity grid switched off the power to about a quarter of a million consumers for several hours in December 2015 (Zetter, 2016).

Finally, there are multiple vulnerabilities arising from *human error and deliberate malfeasance of disgruntled (ex)employees.* Technical exploits can be significantly aided by human error, for example, employees opening phishing emails and installing viruses or malware, or naively inserting infected datasticks into computers (Singer and Friedman, 2014). In other cases, appropriate security software is not installed or is configured incorrectly, or manufacturer installed codes are not changed or system security is not kept up-to-date (Cerrudo, 2015). There are weaknesses in software system designs such that they can be easily and surreptitiously sabotaged by disgruntled present and ex-employees. For example, Goodman (2015) details a case where an ex-employee altered the database records of a vehicle retailer who was using GPS trackers and remote control boxes to re-possess cars, randomly disabling cars and setting off their alarms. In addition, criminal hackers are adept at performing social exploits on trusted employees such as using phishing to release key information (e.g., usernames and passwords) that facilitate access. There is also evidence from the Snowden revelations that "insiders" have been planted by State intelligence agencies with the intention of deliberately compromising the design of networking hardware and fundamental system parameters to facilitate electronic espionage, sabotage, and cyber-warfare (Greenwald, 2014).

These vulnerabilities are exacerbated by a number of factors, not least that it is often unclear as to who is responsible for maintaining security across complex systems and infrastructures when several companies and stakeholders collaborate in their design, supply hardware and software, and operate and use various elements (US DHS, 2016). This is exacerbated with respect to urban management, where city administrations are under increasing pressure for year-on-year "efficiency" savings that affect security in three ways. First, there is long-term under-investment in infrastructure maintenance and an over-reliance on legacy systems. Second, depression of salaries in most public sector organizations make it more difficult to recruit and retain skilled and motivated IT staff to properly implement and maintain smart city technologies. Crucial IT maintenance increasingly uses self-employed contractors and outsourced services, on the one hand deskilling core capacities and eroding institutional memory in the public sector, and on the other creating distributed accountability with a fractured set of bodies (with contracted services, service-level agreements, multi-agencies teams, and remote helpdesks) overseeing security, which often leads to a lack of continuity, coordination, and responsibility. Third, there is a lack of investment in dedicated cybersecurity personnel and leadership (in the form of Chief Information Officer or Chief Technology Officer) and Computer Emergency Response Teams (CERTs) in city governments (Cerrudo, 2015). Cybersecurity expertise is usually limited to a handful of personnel and training across the wider workforce is limited or non-existent (increasing the likelihood of human error). Any cybersecurity plans cities do possess are often siloed with respect to particular systems and departments so that cross-function assessment and response is lacking (Cerrudo, 2015). In addition, it is clear that many smart city vendors have little or no experience in embedding security features into their products—despite claims made in their marketing literature—and many systems possess significant vulnerabilities (Cerrudo, 2015; Lomas, 2015). Furthermore, these vendors can impede security research by limiting access to their systems for testing, thus enabling them to continue to release unsecured products without oversight or accountability (Cerrudo, 2015). Further, too many cities have been lax in insisting on strong security controls and response within the procurement process for new systems.

These five forms of cyberattack and amplifying factors mean that smart city technologies and infrastructures possess a number of security vulnerabilities and risks that are open to being exploited. In the next section, we detail examples of how specific systems have been, or could be, compromised, illustrating the potential scale and impact of such cyberattacks.

Security Vulnerabilities and Risks of Smart Cities

In 2016, the chief information security officer for the city of San Diego government reported that their systems were being hit by an average of 60,000 cyberattacks a day (Anand, 2016). The operators of the electricity supply grid in the United States report being under near constant cyberattack, with one utility recording that it was the target of approximately 10,000 cyberattacks each month (Markey and Waxman, 2013). Indeed, all five commissioners of the Federal Energy Regulatory Commission in the United States agree that sustained and persistent cyberattacks against the digital infrastructure controlling the supply grid is the most serious threat to electricity reliability in American metropolitan areas (Markey and Waxman, 2013). Likewise, the Israel Electric Corp. reports that its servers register about 6,000 unique computer attacks every second, with other critical infrastructure also under continuous attempts to gain access (Paganini, 2013). Many of these cyberattacks are relatively inconsequential, such as randomly directed probes of connected computers and scans across publicly available Internet addresses, and are unsuccessful. However, a small number are much more significant and involve a security breach. Between 2010 and 2014, the US Department of Energy (that oversees the power grid, regulates power generation, and manages the nuclear weapons arsenal) documented 1,131 cyberattacks, of which 159 were successful (Reilly, 2015). In 53 cases, these attacks were "root compromises," meaning that the attackers gained administrative privileges to computer systems, stealing various kinds of personnel and operational information, and potentially doing other damage (Reilly, 2015). In a 2014 study of nearly 600 utility, oil and gas, and manufacturing companies, about 70 percent reported at least one security breach that led to the loss of confidential information or disruption of operations in the previous 12 months (Prince, 2014); 78 percent expected a successful attack on their industrial control systems (ICS) or supervisory control and data acquisition (SCADA) systems in the next two years (Prince, 2014).

Similarly, there have been a number of cyberattacks on transport management systems, as well as proof-of-concept demonstrations of possible attacks. While the idea of crippling a city by disrupting the flow of traffic by hacking its management is not new—for example, it was a central plot device in the 1969 heist movie, *The Italian Job*—it can now be done remotely and is harder to defend against. For example, a cyberattack on a key toll road in Haifa, Israel, closed it for eight hours causing major traffic disruption (Paganini, 2013). A ransomware attack on the San Francisco municipal rail network led to ticketing machines being removed from service for two days (Gibbs, 2016). A research team from the University of Michigan managed to hack and manipulate more than a thousand wireless-accessible traffic signals in one city using a laptop, custom-software, and a directional radio transmitter (Ghena et al., 2014). Likewise, security consultants IOActive Labs have hacked traffic control sensors widely used around the world and altered traffic light sequencing and interactive speed and road signs (Cerrudo, 2014). A teenager in Lodz, Poland, managed to hack the city's tram switches, causing four trams to derail and injuring a number of passengers (Nanni, 2013; Goodman, 2015). In the United States, air traffic control systems have been hacked, Federal Aviation Administration servers compromised, malicious code installed onto control networks, and the personal information of 58,000 workers stolen (Goodman, 2015). Vehicles are also open to being hacked given that a new car contains up to 200 sensors connected to around 40 electronic control units and can connect to wireless networks (Greenburg, 2015).

Indeed, every type of smart city solution and particular system components, including SCADA systems, the sensors and microcontrollers of IoT, and network routers and telecommunication switches, are open to various forms of cyberattack. All essential urban services including the electricity grid, water supply, and road traffic control rely on SCADA systems that are used to control functions and material flows. These systems measure how an infrastructure is performing in real-time and enable either automated or human operator interventions to change settings. The implementation of SCADA systems can be traced back to the 1920s, but were extensively rolled out in the 1980s. As a consequence, many deployments are quite dated and some will contain "forever-day" exploits.[3] A number of SCADA systems have been compromised, with hackers altering how the infrastructure performs, or causing a denial-of-service, or controlling stolen data. The most infamous SCADA hack to date was the 2009 Stuxnet attack on Iran's uranium enrichment plant in which the system was infected by malware that destroyed a number of centrifuges by running them beyond their design specifications. By 2010 over 90,000 Stuxnet infections were reported in 115 countries (Zetter, 2015).

IoT refers to the connecting together of machine-readable, uniquely identifiable objects through the Internet such that they can communicate largely autonomously and automatically. Some objects are passive and can simply be scanned or sensed (such as smart cards with embedded RFID chips[4] used to access buildings and transport systems). Others are more active and include microcontrollers and actuators. All kinds of objects that used to be "dumb," such as thermostats, domestic appliances, security cameras, and lighting systems, are now becoming networked and "smart," generating information about their use and becoming controllable from a distance. The security of IoT is highly variable, with some systems lacking encryption or usernames and passwords, and others open to infection by malware and firmware modification. The complex interdependencies of IoT mean that it has a large attack surface and multiple vulnerabilities (See Table 1). Demonstrating the scale of IoT vulnerability, one provocative project, Insecam.org provides access to the feeds of thousands of unsecured and secured cameras available on the public Internet from cities across the world (Cox, 2014) (See Figure 1). These cameras can also be turned off, with some lacking the function to be restarted remotely (Cerrudo, 2015). Others researchers have shown how to hack into smart lighting and take over control, with potentially serious consequences for personal safety (Chacos, 2016). In addition, IoT infrastructure can be used to perform other kinds of hacks, as

Table 1. The dimensions of risk of Internet of Things technologies

Attack Surface	Vulnerability	Attack Surface	Vulnerability
Ecosystem Access Control	Implicit trust between components Enrolment security Decommissioning system Lost access procedures	**Local Data Storage**	Unencrypted data Data encrypted with discovered keys Lack of data integrity checks
Device Memory	Cleartext usernames Cleartext passwords Third-party credentials Encryption keys	**Third-party Backend APIs**	Unencrypted PII sent Encrypted PII sent Device information leaked Location leaked
Device Physical Interfaces	Firmware extraction User command line interface Administrative command line interface Privilege escalation Reset to insecure state Removal of storage media	**Vendor Backend APIs**	Inherent trust of cloud or mobile application Weak authentication Weak access controls Injection attacks
Device Web Interface	SQL injection Cross-site scripting Cross-site Request Forgery Username enumeration Weak passwords Account lockout Known default credentials	**Update Mechanism**	Update sent without encryption Updates not signed Update location writable Update verification Malicious update Missing update mechanism No manual update mechanism
Device Firmware	Hardcoded credentials Sensitive information disclosure Sensitive URL disclosure Encryption keys Firmware version display and/or last update date	**Ecosystem Communication**	Health checks Heartbeats Ecosystem commands Deprovisioning Pushing updates
Device Network Services	Information disclosure User command line interface Administrative command line interface Injection Denial of Service Unencrypted Services Poorly implemented encryption Test/Development Services Buffer Overflow UPnP Vulnerable UDP Services DoS	**Mobile Application**	Implicitly trusted by device or cloud Username enumeration Account lockout Known default credentials Weak passwords Insecure data storage Transport encryption Insecure password recovery mechanism Two-factor authentication
Administrative Interface	SQL injection Cross-site scripting Cross-site Request Forgery Username enumeration Weak passwords Account lockout Known default credentials Security/encryption options Logging options Two-factor authentication Inability to wipe device	**Cloud Web Interface**	SQL injection Cross-site scripting Cross-site Request Forgery Username enumeration Weak passwords Account lockout Known default credentials Transport encryption Insecure password recovery mechanism Two-factor authentication
Network Traffic	LAN LAN to Internet Short range Non-standard		

Source: Adapted from Open Web Application Security Project, 2015, https://www.owasp.org/index.php/IoT_Attack_Surface_Areas (CC-BY) Accessed November 28, 2017.

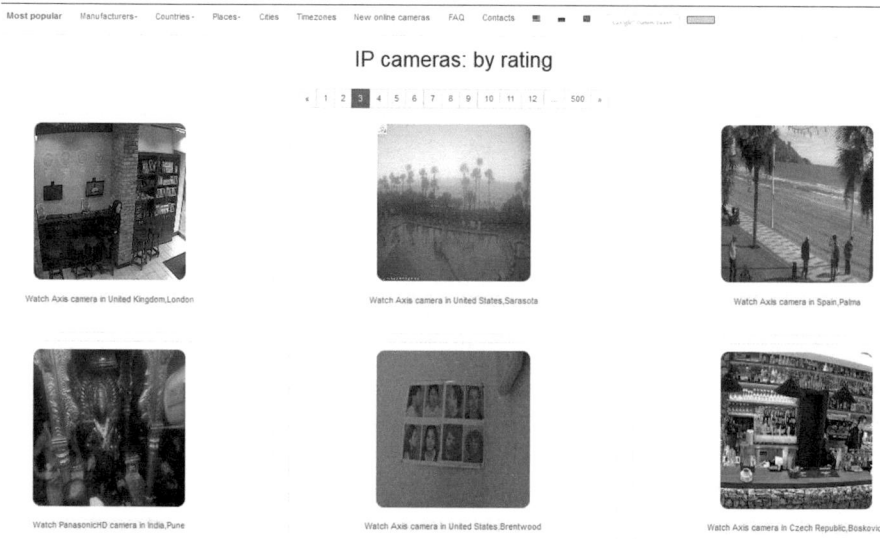

Figure 1. A demonstration of global cybersecurity issues in terms of open security video feeds and webcams on the Internet. Source: Authors' screenshot of *insecam.org*

with the Dyn denial of service attacks in autumn of 2016 in which many significant websites were disrupted by the Mirai botnet that took over unsecured IoT devices and used them to bombard Dyn servers (Woolf, 2016).

Smart city technologies are linked together via a number of communications technologies and protocols such as Long Term Evolution (4G LTE), Global System for Mobile Communication (GSM), Code Division Multiple Access (CDMA), WiFi, bluetooth, Near-Field Communication (NFC), open wireless standard (ZigBee), and wireless communication (Z-Wave). Each of the modes of networking and transferring data are known to have security issues that enable data to be intercepted by third parties and provide unauthorized access to devices. Some of these protocols are so complicated that they are difficult to implement securely. Likewise, telecommunication switches that link together the local and long distance Internet infrastructure are known to have vulnerabilities, including manufacturer and operator back-door security access and access codes that are infrequently updated (Singh and Pelton, 2013). In addition, due to "oversubscribing," wherein wireless carriers want to maximize use of spectrum licenses, networks only have capacity for a fraction of subscribers meaning that during a crisis when demand surges, the system cannot cope, failing to connect both people and things (Townsend, 2013).

What this discussion highlights is the scale and diversity of security flaws in the smart city, their potential vulnerability to cyberattack, and the consequences of such attacks with respect to human safety and social and economic resilience. A key question therefore concerns how such vulnerabilities can be addressed to minimize threats and risk?

Securing Smart Cities: Mitigation and Preventative

To date, the strategy adopted for securing the smart city has largely been one of conventional, largely technical mitigation solutions, such as access controls, encryption, IT industry standards and security protocols, and software patching regimes, along with staff training. While this has had some effect, given the vital nature of smart city technologies and infrastructures to urban life, we contend that securing such systems requires a wider set of systemic interventions that encompass mitigation (lessening the force or intensity of something occurring) and prevention (stopping something from happening or arising), and ensures enactment through both market-led initiatives and governance-led regulation and enforcement.

Conventional Mitigation Solutions

As noted, smart city technologies typically present large attack surfaces that expose a number of potential vulnerabilities, especially in control systems that contain legacy components using old software which has not been regularly patched. The typical approach to securing smart city systems has been to utilize a suite of well-known technical solutions and software security approaches to try and prevent access and to enable restoration if a compromise occurs. For example, the use of access controls (username/password, two-stage authentication, biometric identifiers), properly maintained firewalls, virus and malware checkers, end-to-end strong encryption, and procedures to ensure routine software patching and ability to respond with urgent updates to close exploits as they occur, audit trails of usage and change logs, and effective offsite backups and emergency recovery plans (See Table 2). Using these techniques, the aim is to reduce the attack surface as much as possible and to make the surface that is visible robust and resilient and quickly recoverable in case of failure. Where feasible, systems should have built-in redundancy to ensure that if the primary delivery of a critical system fails, a secondary system automatically takes its place. Such redundancy might include the use of decentralized cloud-based solutions (where data and computation is distributed across sites) or a completely separate technological solution. While an optimal solution, it is also the case that creating genuine redundancy is often difficult and expensive. Indeed, the extent to which the protections detailed in Table 2 are available varies across technologies and vendors; and the application across different institutions and companies is also inconsistent. Moreover, in complex, distributed systems with many components these solutions need to work equally across the complete system since the whole infrastructure/enterprise is only as strong as the weakest link. Further, it is often the case that these kinds of solutions are layered on after a system has been developed rather than being "baked-into" the design.

Table 2. Standard technical aspects of software system security

Access	Updating	Functionality	Design
Effective end-to-end encryption on all communications	Up-to-date virus and malware checkers	Disabling unnecessary functionality	Isolating trusted resources from non-trusted
Enforce strong passwords and access controls	Automatically installing security patch updates on all components, including firmware, software, communications, and interfaces	Ensuring full backup of data and recovery mechanisms	Ensuring that there are no weak links between components
Firewalls			Implementing fail safe and manual overrides on all systems
Audit trails			Ensuring redundancy of systems where feasible

Source: Authors, derived from Martínez-Ballesté et al., 2013 and Cerrudo, 2015

These technical solutions are often bolstered by a vigilant IT staff whose job it is to oversee the day-to-day maintenance of these systems, including monitoring security issues and reacting swiftly to new cyberattacks and breaches. In addition, non IT-staff across an organization can be trained to maintain good practices with respect to security, such as adopting stronger passwords, routinely updating software, encrypting files, and avoiding phishing attacks. However, training is often conducted only once and ongoing staff compliance with best practice is not monitored.

While these security measures have genuine utility, they are far from a complete solution, particularly as smart technologies become ever more critical to the smooth functioning of cities. Instead, a more systemic approach needs to be adopted in relation to both technical design and training. In particular, a *security-by-design* approach that is proactive and preventative, rather than reactive and remedial, needs to be employed by city governments and key institutions responsible for urban management and infrastructure provision. Security-by-design seeks to build strong security measures into systems from the outset rather than attempting to layer them on after initial development. This requires security risk assessment to be a fundamental part of the design process and all aspects of security systems to be rigorously tested before the product is sold (Lomas, 2015); including a pilot phase within a living lab environment that includes testing the security of a product when deployed in real-world contexts and operating as part of a wider network of technologies (to ensure end-to-end security). It also means having in place an ongoing commitment to cybersecurity, including a mechanism to monitor products throughout their life cycle, a process of supporting and patching them over time, and a procedure for notifying customers when security risks are identified. With respect to existing city software systems and control infrastructure, all vendors should be asked for full security documentation and procedures, and a comprehensive testing of their security should be undertaken to identify weak points, undertake remedial security patching, and to upgrade future service level agreements with respect to enhanced security. This is especially the case for legacy systems. If systems cannot be remedially fixed and forever-day exploits remain that could bring down critical systems, then firm plans need to be put in place for upgrades or replacement.

With respect to overseeing the security aspects of smart city technologies we would advocate the formation of a core security team within urban administrations with specialist skills and responsibilities above and beyond day-to-day IT-administration. The work of this team would include: undertaking wide-ranging threat and risk modelling; actively testing the security of smart city technologies (rather than simply monitoring and trusting vendor reassurances); conducting ongoing security assessments; preparing and reviewing detailed plans of action for different kinds of cybersecurity incidents; liaising with the city departments and companies administering smart city initiatives; and coordinating staff training on security issues. The staff would also constitute a city's Computer Emergency Response Team to actively tackle any on-going cybersecurity incidents (Cerrudo, 2015). As a routine part of their work, the core security team should consult with cybersecurity vendors to stay up-to-date on potential threats and solutions (Nanni, 2013). In addition, the team should create a formal channel for security feedback and ethical disclosure, enabling bugs and security weaknesses to be reported by members of the consultants, academics, and allied technology companies. Initial security assessments would be carried out as early as possible, for example in the scoping and procurement phases of technological adoption, to ensure the solutions developed conform to expectations. Part of any assessment should be a consideration of whether systems should be kept in siloes to limit cascade effects. Given cost constraints and lack of strategic foresight, very few cities presently have core security teams or CERTs and are therefore underprepared to deal with a serious cyberattack.

In addition, a step-change in education and training vis-à-vis cybersecurity is required for all those involved in smart city ventures. Within local government and public service/ infrastructure providers, advanced security training should be developed and implemented across the organization, but especially for those involved in the procurement, rollout, and daily running of smart city technologies. This is important because although a system might have an extensive and robust set of technical security solutions these can be nullified by social exploits or human error. Similarly, such programs should be instituted for developers and vendors to stress the need for a security-by-design approach, especially for start-ups and SMEs who might not have the in-house capacity for security expertise. In both cases, training needs to be part of a continual program of professional development to refresh best practice and keep abreast of new technologies and vulnerabilities. We have found very little evidence of such system-wide security training programs other than relatively light introductory courses, often taken on a one-off basis.

Enactment and Enforcement

While it is one thing to advocate for stronger mitigation measures, it is another to ensure that a more systemic approach to cybersecurity for smart cities is widely enacted and enforced. Therefore, there is a need to think about the most appropriate mechanisms to incentivize participation by both the public and commercial sector, and to penalize those who fail to improve security of their products, systems, and services. In general, there are two routes to improving mitigation measures: market-led adoption and government-led regulation and legal enforcement.

The market-led approach consists of vendors developing smart city technologies taking a proactive, self-regulatory stance to security. Here, software companies choose to adopt security-by-design as a de facto standard, collaborate with each other to create effective industry-wide standards, and establish best practices. They ensure security across complex, interdependent systems, and work more closely with the rapidly growing cybersecurity industry in order to improve their products. In so doing, security becomes an expected norm and the adoption of a serious approach to security by companies provides competitive advantage over those that do not comply. In part, the market-led approach would be driven by competition; fear of reputational damage and litigation caused by a major security scandal; and the benefits of self-regulation, rather than the approach of enforcement through legal penalties and fines. While a market-led approach to security does presently exist, it predominately adopts the weak mitigation approach detailed above and not security-by-design. In part this is because there is currently weak pressure from buyers for enhanced security, mainly due to a poor understanding of security vulnerabilities and their potential consequences and inadequate procurement practices. Moreover, the imperatives to get product to market as quickly as possible (often to pre-empt a competitor) and turn a profit mean that security corners are being cut. As such, market-led responses should be accompanied by more "top-down" regulation and better management practices by city authorities and urban infrastructure operators.

The regulation and management-led approach seeks to encourage secure deployment of smart city technologies through compliance measures and active oversight. The former requires the formulation of security standards, directives, and best practices that smart city deployments must comply with or face some form of penalty, such as prosecution, fines, or loss of contract. There are now a host of smart city standards initiatives underway—by bodies such as the International Standards Organization, British Standards Institute, American National Standards Institute, City Protocol—aimed at defining minimum specifications for technical development and deployment of core technologies. The latter necessitates setting up management structures and procedures for ensuring compliance is being met and enforced. For example, large public bodies that are operating the European Union have to institute an audit and risk committee that identifies vulnerabilities and monitors potential threats to an organization and oversees mitigation strategies. These are often broad in scope and could benefit from a sub-committee focused specifically on software security and network threats. This sub-committee should oversee and audit the work of the core security team; advise on the work priorities and program; certify security assessments; certify that the city's smart city technologies conform to legal and regulatory requirements; ensure that response and mitigation plans and processes are in place; and ensure there is clear communication to the public concerning the security of smart city systems (Nanni, 2013). In addition, city administrations should bake security-by-design and on-going security maintenance (including on-time patching and 24/7 incident response) into the procurement process and subsequent service level contracts, with the extent to which the proposed solutions meet desired parameters directly influencing the evaluation of tenders (Cerrudo, 2015). They should also support whistleblowers who wish to expose security vulnerabilities and require the public reporting of security breaches.

We have found no example of a city that presently enacts such systemic, enhanced security oversight or procurement beyond seeking existing mitigation strategies. For the most part this is due to a lack of in-depth knowledge and competence, and institutional inertia. Consequently, smart city technologies have been in the past and are still being procured with little coordinated consideration of security harms and slotted into existing city management in an ad hoc fashion with minimal strategic foresight. Given the potential harms and the associated costs that can arise, this piecemeal and make-do approach needs to be discontinued to be replaced with a more systemic and coordinated approach.

A Preventative Approach

Even with a strong mitigation strategy and effective enforcement procedures, it is not possible to eradicate all the security vulnerabilities and associated risks from the smart city. There is, therefore, a case to be made for considering a preventative approach, one that involves building some urban infrastructure and control systems that are deliberately "deaf" (not networked and remotely accessible) and "dumb" (i.e., not automated by code), which would elide many software security overheads. A preventative approach is quite straightforward to articulate—simply put, "do not adopt smart city technologies as presently conceived;" the best way to prevent risks from materializing is not to create vulnerabilities in the first place.

Yet making the case for such an approach is much more difficult in practice because of the perceived benefits of creating a smart city. Such a cautious, preventative approach, that questions seriously the commercial logics and profit streams of many hardware vendors and software developers, will be labelled "backward looking" and "out-of-date," and derided for having a neo-Luddite mentality (See Jones, 2013). Indeed, at present, advocating a preventative approach would be considered a radical means of securing smart cities as it requires a reframing of the value around technology and a rethinking of the balance between convenience/efficiency and security/safety. It requires a counter-narrative against "smarter is better" and advocacy for conventional electro-mechanical components and systems that run reliably without additional software monitoring and network access.

There is a case, however, to be made that the potential risks networked infrastructure pose, plus the cybersecurity, management and training costs of ensuring security, outweigh the efficiency and functionality gains promised and the "inconveniences" of maintaining "air-gapped" technologies (that is, systems that are physically isolated from other networks). Having to send a person physically to a component to re-activate it, reconfigure settings, or repair it might seem costly and burdensome when it could be done remotely. Indeed, in the era of ubiquitous connectivity, cloud-computing, integrated and interoperable systems, and remote control, the notion of having an air gap in critical systems might seem counter-intuitive. However, it can be an effective method of security that prevents hacking and cascade effects and significantly reduces vulnerabilities.

Equally, there are reasons to be skeptical of the benefits claimed by advocates (who are often self-interested) for new cyber-physical systems as it is well noted that they tend to oversell the promises of smart city technologies while ignoring their perils (Townsend, 2013; Greenfield, 2013; Kitchin, 2014; Datta, 2015). Certainly, many existing smart city system deployments have not delivered the anticipated gains in efficiency, flexibility, productivity and convenience; in many cases, especially with regards to the IoT, objects and systems have been digitally networked for little perceivable gain or real benefits to the functioning and management of cities (though they benefit vendors through their sale/servicing and potential monetization of data streams). In fact, if anything, some newly software-enabled systems make routine tasks more complex to complete, error-prone, unreliable, stressful, costly in time and cognitive attention, and less secure, as well as raising issues with respect to excessive surveillance and privacy (Greenfield, 2013; Kitchin, 2016). In other words, networking city infrastructure and introducing new systems do not necessarily improve performance, yet they do make them more vulnerable to security risks.

Nonetheless, at present, implementing preventative measures will be difficult to promote and promulgate given the widespread adoption of techno-utopian discourses of "progress" enacted by smart urbanism. This is especially the case in the current neoliberal climate that encourages cities to form public-private partnerships with companies and to outsource or privatize services, and where access to government grants will be difficult without claiming to create and implement innovative and cutting-edge smart city solutions. This may change though if the "cutting-edge" of city management becomes recognized as the "bleeding-edge" of insecurity.

Conclusion

In this paper we have examined in-depth the current state of play with regards to the security of smart cities. In an ironic twist, smart city technologies are promoted as an effective way to counter and manage uncertainty and risk in present-day cities, yet they paradoxically create new risks, including making city infrastructure and services insecure, brittle, and open to extensive forms of vandalism, disruption and criminal exploitation. This paradox has largely been ignored by commercial and governmental interests or tackled through a traditional mitigation approach. This is perhaps no surprise. Although we have identified five forms of vulnerability and detailed the present extent of cyberattacks on city infrastructure and services, presenting a number of illustrative examples where they have been compromised, as far as is publicly known, the majority of attacks are presently being repulsed using cybersecurity software tools and management practices, or their effects have been only locally disruptive or damaging but not critical for the long-term delivery of services (Singer and Friedman, 2014). Indeed, despite widespread, low-level attempts, successful cyberattacks on city systems are still relatively rare and when they have occurred their effects generally last no more than a few hours or involve the theft of data rather than creating life-threatening situations. That said, even short-term disturbances, such as shutting down the electricity grid for a few hours or causing traffic gridlock, can be an expensive disruption through lost productivity and opportunities, and can also be potentially life-threatening. They also signal the threat of more damaging cyberattacks in coming years as actors develop more sophisticated methods of hacking, and security measures fail to keep pace.

Indeed, it is clear that smart city technologies presently have multiple vulnerabilities and that these are and will be exploited for various ends. Moreover, there is a cybersecurity "arms race" underway between attackers and defenders, and it may be that more severe disruption of critical infrastructure has so far been avoided because nation-state actors do not want to reveal their capabilities and they fear retaliation from adversaries (Rainie et al., 2014). In other words, smart city technologies are vulnerable to cyberattack and cyberterrorism and existing vulnerabilities are only likely to increase in the future. In our view, present strategies for addressing the vulnerabilities and risks posed by the mass adoption of networked technologies for city management are woefully inadequate and predominantly rely on existing technical and training mitigation strategies and market-led solutions.

Instead, we advocate a widening and deepening of mitigation strategies to include security-by-design as a de facto approach for all future smart city procurement, a comprehensive assessment of existing urban infrastructures and information systems and remedial security patching or replacement, the formation of core security and computer emergency response teams within city administrations with specialist skills and responsibilities beyond general IT-administration, and a step-change in security training and continuing professional development in both public and commercial sectors. This should be complemented by a management and regulation approach to smart city technologies and implementation, rather than simply a market-led approach, to ensure active oversight and compliance with security standards, best practices, municipal policy, and third-party service contracts. We also suggest that serious consideration be given to a preventative approach to security, wherein critical infrastructure is air-gapped or not given the "smart" treatment when it is not really needed. In addition, the use of new technical solutions to access, control, and authentication, such as the use of blockchains, warrant examination (see Christidis and Devetsikiotis, 2016).

It is self-evidently too late to roll-back the smart city agenda, and much of the adoption of smart city technologies by municipal authorities across the world cannot simply be removed. However, it is not too late to recognize the extent of the new security vulnerabilities, threats, and risks posed by these technologies and to put in place strategies and approaches to mitigate and prevent them. We believe that not enough is presently being done by vendors and city administrators to identify vulnerabilities and risks and to formulate effective responses. Moreover, infrastructural and city system security issues are largely being ignored within the social sciences and urban studies, the consequence of which is to leave the formulation of solutions largely to computer and engineering sciences and market-solutions which favor technical and for-profit approaches rather than more policy, governance, and for-public-good tactics. Securing smart cities requires a holistic understanding of a city as a diverse, contested set of places, rather than constituting a system of systems. It necessitates a comprehension of the reasons why vulnerabilities exist and how they might be tackled beyond technical fixes; the potential social and economic consequences of risks being realized; and how risks and potential solutions are contextualized within the acting regime and political economy of urban development and management. While we have sought to examine the security of smart cities from a normative perspective, this undoubtedly needs to be complemented with a wider social, political, and economic analysis of security issues and smart urbanism.

Notes

1. Computer hacking culture has a long history and with diverse and contested meanings (Levy, 1984), but the term has come typically to be applied to those with malicious or criminal intent.
2. The IoT is a fast-developing set of identification and technologies that connect together formally "dumb" physical objects and make them addressable through the internet and potential facilitate all manner of new activities and processes in relation to these objects, often in highly automated and autonomous fashion. As such, IoT is the critical element in the creation of what Dodge and Kitchin have called the "machine-readable world."
3. This is a publicly known code error in a software product that the vendor is not able to or is not intending to fix, consequently there is no means of patching the software.
4. These comprise a small physical electronic circuit and antenna that can be fix to physical objects and automatically broadcast a globally unique identification code number when queried by appropriate radio signal: See Frith, 2015.

Disclosure Statement

No potential conflict of interest was reported by the authors.

Funding

The research funding for this paper was provided by a European Research Council Advanced Investigator Award, 'The Programmable City' [no. ERC-2012-AdG-323636] and the Department of the Taoiseach, Ireland.

ORCID

Rob Kitchin ⓘ http://orcid.org/0000-0003-4458-7299

Bibliography

P. Anand, "The 'Mind-Boggling' Risks your City Faces from Cyber Attackers," *Market Watch* (30 January 2016) <www.marketwatch.com/story/the-mind-boggling-risks-your-city-faces-from-cyber-attackers-2016-01-04> Accessed October 5, 2017.

Article 29 DPWP, *Opinion 8/2014 on the Recent Developments on the Internet of Things* (Article 29 Data Protection Working Party, 2014) <http://ec.europa.eu/justice/data-protection/article-29/documentation/opinion-recommendation/files/2014/wp223_en.pdf> Accessed October 5, 2017.

U. Beck, *Risk Society: Towards a New Modernity* (London: Sage, 1992).

S.M. Bellovin, "Attack Surfaces," *IEEE Security and Privacy* 14: 3 (2016) 88.

R. Bodenheim, J. Butts, S. Dunlap, and B. Mullins, "Evaluation of the Ability of the Shodan Search Engine to Identify Internet-facing Industrial Control Devices," *International Journal of Critical Infrastructure Protection* 7 (2014) 114–123.

C. Cerrudo, "Hacking US (and UK, Australia, France, etc.) Traffic Control Systems," *IOActive Blog*, (30 April, 2014) <http://blog.ioactive.com/2014/04/hacking-us-and-uk-australia-france-etc.html> Accessed October 5, 2017.

C. Cerrudo, "An Emerging US (and World) Threat: Cities Wide Open to Cyber Attacks," *Securing Smart Cities* (2015) <http://securingsmartcities.org/wp-content/uploads/2015/05/CitiesWideOpen ToCyberAttacks.pdf> Accessed October 5, 2017.

B. Chacos, "Osram's Lightify Smart Bulbs Suffer from Several Serious Security Flaws," *PC World* (27 July 2016) <www.pcworld.com/article/3101008/connected-home/osrams-lightify-smart-bulbs-suffer-from-several-serious-security-flaws.html> Accessed October 5, 2017.

K. Christidis and M. Devetsikiotis, "Blockchains and Smart Contracts for the Internet of Things," *IEEE Access* 4 (2016) 2292–2303.

J. Cox, "This Website Streams Camera Footage from Users Who Didn't Change Their Password," *Motherboard* (31 October 2014) <http://motherboard.vice.com/read/this-website-streams-camera-footage-from-users-who-didnt-change-their-password> Accessed October 5, 2017.

A. Datta, "New Urban Utopias of Postcolonial India: 'Entrepreneurial Urbanization' in Dholera Smart City, Gujarat," *Dialogues in Human Geography* 5: 1 (2015) 3–22.

S. Durbin, "Building Smart City Security," *TechCrunch* (12 September 2015) <www.techcrunch.com/2015/09/12/building-smart-city-security> Accessed October 5, 2017.

D.J. Evans and D.T. Herbert, *Geography of Crime* (London: Routledge, 1989).

J. Frith, "Communicating Behind the Scenes: A Primer on Radio Frequency Identification (RFID)," *Mobile Media and Communication* 3: 1 (2015) 91–105.

B. Ghena, W. Beyer, A. Hillaker, J. Pevarnek, and J.A. Halderman, "Green Lights Forever: Analyzing the Security of Traffic Infrastructure," *Proceedings of the 8th USENIX Workshop on Offensive Technologies* (2014) <www.usenix.org/system/files/conference/woot14/woot14-ghena.pdf> Accessed October 5, 2017.

S. Gibbs, "Ransomware Attack on San Francisco Public transit Gives Everyone a Free Ride," *Guardian* (28 November 2016) <www.theguardian.com/technology/2016/nov/28/passengers-free-ride-san-francisco-muni-ransomeware> Accessed October 5, 2017.

M. Goodman, *Future Crimes* (New York: Bantam Press, 2015).

A. Greenburg, "Hackers Remotely Kill a Jeep on the Highway—With Me in It," *Wired* (21 July 2015) <www.wired.com/2015/07/hackers-remotely-kill-jeep-highway> Accessed October 5, 2017.

A. Greenfield, *Against the Smart City* (New York: Do, 2013).

G. Greenwald, *No Place to Hide: Edward Snowden, the NSA, and the US Surveillance State* (New York: Macmillan, 2014).

T. Hall, "Geographies of the Illicit: Globalization and Organized Crime," *Progress in Human Geography* 37: 3 (2013) 366–685.

A. Hern, "Ransomware Threat on the Rise," *Guardian* (3 August 2016) <www.theguardian.com/technology/2016/aug/03/ransomware-threat-on-the-rise-as-40-of-businesses-attacked> Accessed October 5, 2017.

S.E. Jones, *Against Technology: From the Luddites to Neo-Luddism* (London: Routledge, 2013).

R. Kitchin, "The Real-Time City? Big Data and Smart Urbanism," *GeoJournal* 79: 1 (2014) 1–14.

R. Kitchin, "The Ethics of Smart Cities and Urban Science," *Philosophical Transactions A* 374: 2083 (2016) 1–15.

R. Kitchin and M. Dodge, *Code/Space: Software and Everyday Life* (Cambridge: MIT Press, 2011).

J.L. LeBeau and M. Leitner, "Progress in Research on the Geography of Crime," *The Professional Geographer* 63: 2 (2011) 161–173.

S. Levy, *Hackers: Heroes of the Computer Revolution* (Harmondsworth: Penguin, 1984).

P.L. Li, M. Shaw, J. Herbsleb, B. Ray, and P. Santhanam, "Empirical Evaluation of Defect Projection Models for Widely-deployed Production Software Systems," *ACM SIGSOFT Software Engineering Notes* 29: 6 (2004) 263–272.

R.G. Little, "Managing the Risk of Cascading Failure in Complex Urban Infrastructures," in S. Graham, ed., *Disrupted Cities: When Infrastructure Fails* (London: Routledge, 2010) pp. 27–39.

N. Lomas, "The FTC Warns Internet of Things Businesses to Bake in Privacy and Security," *TechCrunch* (8 January 2015) <http://techcrunch.com/2015/01/08/ftc-iot-privacy-warning> Accessed October 5, 2017.

A. Luque-Ayala and S. Marvin, "The Maintenance of Urban Circulation: An Operational Logic of Infrastructural Control," *Environment and Planning D: Society and Space* 34: 2 (2016) 191–208.

D. MacKinnon and K.D. Derickson, "From Resilience to Resourcefulness: A Critique of Resilience Policy and Activism," *Progress in Human Geography* 37: 2 (2013) 253–270.

G. Manaugh, *A Burglar's Guide to the City* (New York: Farrar, Straus & Giroux, 2016).

E.J. Markey and H.A. Waxman, "Electric Grid Vulnerability: Industry Response Reveal Security Gaps" (2013) <www.markey.senate.gov/imo/media/doc/Markey%20Grid%20Report_05.21.131.pdf > Accessed October 5, 2017.

A. Martínez-Ballesté, P.A. Pérez-Martínez, and A. Solanas, "The Pursuit of Citizens' Privacy: A Privacy-Aware Smart City is Possible," *IEEE Communications Magazine* 51: 6 (2013) 136–141.

G. Nanni, *Transformational "Smart Cities": Cyber Security and Resilience* (Mountain View, CA: Symantec, 2013).

W.A. Owens, K.W. Dam, and H.S. Lin, *Technology, Policy, Law, and Ethics Regarding US Acquisition and Use of Cyberattack Capabilities* (Washington DC: Committee on Offensive Information Warfare; National Research Council, National Academic Press, 2009).

P. Paganini, "Israeli Road Control System hacked, Caused Traffic Jam on Haifa Highway," *Hacker News* (28 October 2013) <http://thehackernews.com/2013/10/israeli-road-control-system-hacked.html> Accessed October 5, 2017.

C. Perrow, *Normal Accidents: Living With High-Risk Technologies* (New York: Basic Books, 1984).

B. Prince, "Almost 70 Percent of Critical Infrastructure Companies Breached in Last 12 Months: Survey," *Security Week* (14 July 2014) <www.securityweek.com/almost-70-percent-critical-infrastructure-companies-breached-last-12-months-survey> Accessed October 5, 2017.

L. Rainie, J. Anders, and J. Connolly, "Cyber Attacks Likely to Increase," *Digital Life in 2025* (Pew Research Center, 2014) <www.pewinternet.org/files/2014/10/PI_FutureofCyberattacks_102914_pdf.pdf> Accessed October 5, 2017.

S. Reilly, "Records: Energy Department Struck by Cyber Attacks," *USA Today* (11 September 2015) <www.usatoday.com/story/news/2015/09/09/cyber-attacks-doe-energy/71929786/>   Accessed October 5, 2017.

S. Sarma, "I Helped Invent the Internet of Things: Here's Why I'm Worried about How Secure It Is," *Politico* (June 2015) <www.politico.com/agenda/story/2015/06/internet-of-things-privacy-risks-security-000096> Accessed October 5, 2017.

B. Schneier, *Beyond Fear: Thinking Sensibly about Security in an Uncertain World* (New York: Copernicus Books, 2003).

P.W. Singer and A. Friedman, *Cybersecurity and Cyberwar* (Oxford: Oxford University Press, 2014).

I.B. Singh and J.N. Pelton, "Securing the Cyber City of the Future," *The Futurist* 47: 6 (2013) 22.

O. Söderström, P. Till, and F. Klauser, "Smart Cities as Corporate Storytelling," *City* 18: 3 (2014) 307–320.

A.M. Townsend, *Smart Cities: Big Data, Civic Hackers and the Quest for a New Utopia* (New York: Norton, 2013).

US DHS (United States Department of Homeland Security), *Strategic Principles for Securing the Internet of Things* (15 November 2016) <www.dhs.gov/sites/default/files/publications/Strategic_Principles_for_Securing_the_Internet_of_Things-2016-1115-FINAL.pdf>   Accessed October 5, 2017.

J.M. White, "Anticipatory Logics of the Smart City's Global Imaginary," *Urban Geography* 37: 4 (2016) 572–589.

N. Woolf, "DDoS Attack that disrupted Internet was Largest of its Kind in History," *Guardian* (26 October 2016) <www.theguardian.com/technology/2016/oct/26/ddos-attack-dyn-mirai-botnet> Accessed October 5, 2017.

K. Zetter, *Countdown to Zero Day: Stuxnet and the Launch of the World's First Digital Weapon* (New York: Broadway Books, 2015).

K. Zetter, "Inside the Cunning, Unprecedented Hack of Ukraine's Power Grid," *Wired News* (3 March 2016) <www.wired.com/2016/03/inside-cunning-unprecedented-hack-ukraines-power-grid> Accessed October 5, 2017.

E-Capital and Economic Growth in European Metropolitan Areas: Applying Social Media Messaging in Technology-Based Urban Analysis

Juho Kiuru and Tommi Inkinen ⓘ

ABSTRACT

Innovation is an elemental part of regional economic growth. In the past years, information and communication technologies (ICTs) have enabled new means for data collection, and analysis for the study of regional innovation systems. This paper investigates innovation and technology messaging in Twitter, which has been described as the SMS of the Internet. The concept of electronic capital (e-capital) is applied in order to find out how technology messaging relates to the economic situation in metropolitan areas. The recently introduced concept of e-capital is cultivated from the conceptualizations of innovation acknowledging that different forms of capital, including human, social, and economic, circulate and have an effect on each other. The analysis indicates that clusters of e-capital and potential growth clusters are identifiable by using Twitter activity. In Europe, e-capital agglomerates to previously identified clusters of the "Blue Banana" and the "Golden Banana" (or the "Sun Belt"). Based on spatial statistics, we apply Categories of Metropolitan Areas (COMAs) in order to classify Twitter intensive locations across Europe. We defined four COMAs and estimated their e-capital potential. The most problematic COMA lies in Eastern Europe whereas the strongest concentration is found in Western Europe.

Introduction

Bourdieu (1986) defined social capital as social relationships that are "usable" in the short or long term. Social capital has also been described as "connections among individuals— social networks and the norms of reciprocity and trustworthiness that arise from them" (Putnam, 2000: 19) and "that enable participants to act more effectively to pursue shared objectives" (Putnam, 1995: 664). "Better-connected" people have informational advantages as they gain timely access to high-quality and fine-grained information faster and earlier than "less-connected" people (Burt, 1992; Podolny, 1993). People with more social capital receive more employment information via informal social interactions (Granovetter, 1973; Song, 2013). The information advantage enabled by social capital can translate into higher compensation, faster promotion, and better ideas. The literature on social capital indicates that diverse network contacts require and facilitate a diverse repertoire of cultural knowledge (Erickson, 1996).

There are several studies that can prove the correlation between social capital and regional (e.g., Helliwell and Putnam, 1995; Iyer et al., 2005) and national (Knack and Keefer, 1997; Whiteley, 2000; Woolcock, 2001) economic development. A significant strand of research has measured innovative networks that produce social capital and thus economic growth. Miguelez and Moreno (2013) have investigated the spatial correlation between regional research networks and regional (economic) development in Europe. They concluded that collaborations with inventors outside the region (weak or distant ties) are in fact more important for innovative growth than networks within region (same notion as Granovetter, 1973). In addition, innovative networks between firms are crucial for the growth of the technology industry (e.g., Tsai, 2009) and for the creation of new innovations (Ala-Rämi and Inkinen, 2008). However, there are different empirical findings of the importance of technology collaboration networks (TCNs) between (SME) firms (Fernández-Olmos and Ramírez-Alesón, 2017). There is empiricism of positive correlation between collaboration of firms (e.g., Robson and Bennett, 2000), but also negative correlation (e.g., Nieto and Santamaría, 2007) or non-significant statistic results (e.g., Bougrain and Haudeville, 2002). A recent empirical study from Spain shows positive correlation between TCNs and the sales of new products for SMEs (Fernández-Olmos and Ramírez-Alesón, 2017). However, Franco and Haase (2015) suggest that TCNs between SMEs are under-studied. Thus there are differences in how deep and close the collaboration is and in what stage it is (Fernández-Olmos and Ramírez-Alesón, 2017).

One of the motivators for this study is a result from the Global Innovation Index (Dutta et al., 2017) stating that the primary option for economically sustainable growth for small (regional and national) economies is their openness. Cooke (2017) has recognized that distinctive territorial innovation systems (TIS) have been eclipsed, because they have not adapted to the Global Innovation Network (GIN), which has changed dramatically since the early part of this decade with "new combinations" of commercial modularization (e.g., wireless radio, camera, music, video, film, computing, Internet, and integrative systems design). This study investigates the effects of these developments by studying innovation messaging in European metropolises. The flows associated with this messaging may be captured in digital domains as this study demonstrates.

The utilization of social media databases is still quite a rare method in urban studies (e.g., Cranshaw et al., 2012; Hasan et al., 2013; Hiippala et al., 2018). These studies indicated to us that instead of using methods such as target interviews or surveys, we could find spatially significant data with a greater sample size if the data were to be collected straight from the database of a Social Network Site (SNS) such as Twitter. Thus, tweets (Twitter short messages) are suggested here as a potential proxy for e-capital. This leads us to a concept of "digital social capital" (Mandarano et al., 2010). In this regard, Merisalo (2016: 12) applies the concept of electronic capital, where

the perspective is widened to include not only resources and assets but the entire "process" (Harvey, 2010: 40) of how investments in digitalization produce e-capital that results in added value, which has a potential to convert to other forms of capital.

Merisalo redefines Hall's (2000) original definition of e-capital, which related to college graduate workers' use of business methods based on computers. E-capital is defined as a form of capital, which: "emerges from the possibilities, capabilities, and willingness of individuals, organizations, and societies to invest in, utilize, and reap benefits from digitalization and thus create added value" (Merisalo, 2016: 22). Further, all forms of capital are required, but also produced in the process of e-capital. This means that e-capital is likely to emerge in same locations as other forms of capital. On the other hand, regions "can gain access to other forms of capital by investing and utilizing digitalization, simply by jumping into the e-capital conversation process" (Merisalo, 2016: 32).

Formulating Empirics

The research questions for this paper are:

(1) Does e-capital (approximated through Twitter activity) relate to economic development?
(2) In which metropolitan areas does e-capital concentrate in Europe?
(3) Do the metropolitan areas with high e-capital concentrations correspond to metropolitan areas of high economic development levels?

These questions are derived from the earlier literature. E-capital, like other forms of capital, is likely to agglomerate in cities (Merisalo, 2016). Also, it has been argued that economic growth takes place in clusters (Porter, 2000), in regional innovation systems (RIS) (Cooke 1992), in territorial innovation systems (TIS) (Cooke, 2017), and in urban environments (Rossi, 2016). Therefore, the empirical part of this paper investigates levels of e-capital in cities and in particular in European metropolitan areas (representing regional innovation systems: Cooke 1992), which have been defined as urban areas in Europe (European Commission, 2016a). We are going to form clusters (Porter, 2000) and urban classifications demonstrating the interlinkage of e-capital and economic indicators. With spatial statistics, we advance the familiar used concept of clusters into the broader category of COMAs that do not necessarily locate geographically close to each other.

The approach is quantitative and the interpretation of e-capital relies on the number of tweets according to their presence in European cities. Twitter is an American online news and social networking service in which the user posts and interacts with messages known as "tweets." Both individuals and companies can hold a Twitter account, so we believe we can capture the innovative "buzz" between both individuals and technology-oriented firms. The most likely receiver of a tweet is the one who follow one's tweets, but also everyone interested in these specific hashtags are able to read them. The most likely followers are users from the same region, but also users from anywhere interested in the same themes. As a result, data mining from Twitter API provides an approximation of linkages connecting economic indicators with social media use.

The paper focuses on the presence of tweets depicting innovative argumentation forming social capital. Therefore, data mining tweets contains hashtags related to innovativeness. The empirical design applies keywords identified in the earlier literature. These are "innovation," "startup," "tech," and positive associations to migration and openness. The analysis considers current topics in social media that are not location bound. For example, discussions concerning Brexit (British resignation from the EU) in Twitter seemed to cluster heavily in Britain. An opposite example is the European football championships held in 2016 in France that experienced Tweet accumulation in participating countries but not so much in those countries that did not participate. All in all, hashtags that were data mined from Twitter were "innovation," "startup," and "tech," which are the most used hashtags related to innovation (CyBranding Ltd., 2016), and as a soft factor, "refugeeswelcome," which has been a popular hashtag in Twitter during the recent refugee crisis in Europe and represents in our study tolerance that has been related to innovativeness (Florida, 2002, 2012). The importance of tolerance has been under severe criticism (e.g., Clark, 2003; Glaeser et al., 2004), but positive correlations between tolerance and innovativeness in several studies from other researchers as well (e.g., Niebuhr, 2010; Qian, 2013; Kiuru and Inkinen, 2017) motivated us to test this phenomenon related to innovativeness. Remembering the essential role of openness in regional growth (Dutta et al., 2017), and the fact that several terms including diversity, tolerance, and openness are often interchangeably applied (Qian, 2013), motivates us to take the view of tolerance into an account complementing the hard factors such as "innovation," "startup," and "tech."

The number of tweets presenting the level of e-capital of the metropolitan areas was compared to the level of economic development (GDP) and employment in order to find out whether or not Twitter activity has anything to do with them (research question 1). The first interpretation is that e-capital generates economic capital, as Merisalo (2016) suggests. With economic capital, we mean the economic value the non-economic Bourdieuan capitals create. In this paper, economic indicators are represented with absolute GDP and GDP per capita figures as well as total employment and relative employment rates. The second research question of the paper is to answer where the e-capital has clustered in Europe and what are the potential growth areas regarding the gap between e-capital and economic capital. The second interpretation concerns the school of thought which considers digitalization in the light of inequality and the digital divide (e.g., Chen and Wellman, 2004; Chen, 2013).

In summary, Chen (2013) introduces the dynamics of the digital divide with three examples. First, it has been recognized that technology users (particularly the pioneers or early adopters) are prone to be relatively young, they have a high level of formal education, and they tend to be located in urban environments (Chen et al., 2002; Boase, 2010). Second, one's socioeconomic condition has an effect on one's Internet use skills and to Internet behavior in general (Zillien and Hargittai, 2009). And third, LaRose et al. (2008) indicate that in rural areas there are considerable obstacles related to the expansion of Internet use. These include slow investment levels, problems of access, and the lack of infrastructure. The third research question tackles these observations as it focuses on COMAs, i.e., whether or not e-capital clusters in the areas that have higher economic development levels. The classification of the identified COMAs is presented in the results.

Data and Methods

Economic Data

The main data source is Eurostat, which provides open data from the metropolitan areas in Europe (European Commission, 2016b). Eurostat formed 274 metropolitan areas from NUTS3 regions using Urban Audits definition of Functional Urban Area (FUA) (European Commission, 2016b). There are numerous statistics available from these areas. The selected variables include total population and socioeconomic factors such as GDP (total) and employment (total) (See descriptive statistics in Table 1). We formed also proportional measures including GDP per capita and employment rate. There were some gaps in the data as some of the metropolitan areas missed socioeconomic data. The decision was made to exclude these areas from further analyses. Another reduction came from the languages spoken in European cities. We decided to exclude cities from the United Kingdom in order to avoid a possible bias resulting from the language. Overall, the total sample size includes 234 urban areas. All the data and software used in this study are open access.

Twitter Data

The second data source was Twitter social media site's extensive database. It is available via Twitter API (developer database). The data were processed with an open source programming software R. Geoff Jentry made it possible to connect Twitter API with R open source software. In practice, the data was downloaded in the form of needed packages from *GitHub*. Twitter allows search of tweets from the past two weeks. It is possible to search tweets containing selected keywords as is done here. In addition, it is possible to search tweets from identified locations. This enabled the collection of tweets from every metropolitan area having the needed socioeconomic data from Eurostat. Tweets are thus seen as a proxy to indicate innovation interest (tweeting activity) as a potential for economic growth. Studying tweets, it is also possible to propose an interpretation concerning technology-based urban development. The keywords applied in this analysis were all in English. It is considered that the selected keywords are related to innovativeness and can be applied. Again, these are "innovation," "tech," "startup," and "refugeeswelcome." They all are highly popular hashtags in Twitter (CyBranding, 2016).

The data were extracted with a definition code that produced a list of tweets containing specific hashtags for specific locations for the period of two weeks. For example, the code for innovation tweets from Paris was the following: *searchTwitter ('innovation', n = 30,000, geocode='48.8566, 2.3509, 30mi')*. Where n = expected amount of tweets from last two weeks. The smaller the number of tweets, the faster the software mines the data. Therefore, for smaller metropolises we used n = 1,000 and if command brought us 1,000 tweets, the

Table 1. Descriptive statistics of the used variables from 234 metropolitan areas in Europe

	Mean	Std. Deviation	N
GDP per capita	28,115	12,554	234
Employment rate	46.2	8.2	234
Employment total	501,580	592,361	234
Population	1,072,912	1,184,325	234

limit was raised accordingly to the level that all tweets were obtained. With the largest metropolises, it was necessary to use n = 30,000, making the data mining time consuming. Finally, the geocode is the latitude and the longitude of the location where the tweets are mined from. Coordinates have to be in four-digit form. Coordinates were mined one by one from the website "Find latitude and longitude" (Zwiefelhofer, 2016). After the coordinates, the radius where the tweets are mined can be defined. Radius has to be defined in miles. The decision was made to use a 30-mile radius as it would be a grounded estimate of the size of a European metropolitan area. This is a researcher-dependent decision made in the study, but it is considered to be appropriate and it does not change the interpretations. By changing the parameters to code, all aspects defined in the research questions were processed and the end result is a list of tweets, city by city. The number of tweets was merged into a table including all 234 metropolitan areas having the statistical data of GDP per capita, employment rate, and population from the Eurostat database.

After collecting the number of tweets in metropolitan areas defined by Eurostat, the number of tweets per capita was calculated. The number of tweets is often very small compared to the population of cities so it is more feasible to calculate tweets per 1,000 residents. A regression analysis was made by applying the number of tweets as predictor and GDP and the employment rate as constant variables. With the coefficients of the analysis, it is possible to answer the question regarding which hashtags are in relationship with economic indicators. Descriptive statistics of the data mined from Twitter can be seen in Table 2. The most usual case is tweets containing the work "tech" with 319 on average, in the metropolitan area. The data were collected in two time periods (April 18–May 2, 2016 and July 11–July 25, 2016).

Data Limitations

As discussed, there are limitations related to this approach. First of all, tweets are only from a period of two weeks. This makes single events or users possible to bias the results. While a longer period of examination may have proved steadier results, this was the only option available to us as we collected data from Twitter API. In future research applying similar methods, tweets could be mined from different time periods in order to increase reliability concerning the timing issues in the data collection.

Another debatable selection is the definition of the geographical area. Searching tweets within a 30-mile radius of the city center is an analytical restriction. For example, in coastal metropolitan areas, the area from which the tweets are collected is much smaller than in continental areas. In addition, defining the metropolitan area is a very problematic issue,

Table 2. Descriptive statistics of studied tweet keywords in 234 metropolitan areas

	Mean	Std. Deviation	N
Innovation	241.32	1037.79	234
Startup	260.72	906.49	234
Refugeeswelcome	19.27	59.56	234
Tech	318.50	1506.87	234
Innovation per 1,000 capita	0.15	0.33	234
Startup per 1,000 capita	0.17	0.28	234
Refugeeswelcome per 1,000 capita	0.021	0.076	234
Tech per 1,000 capita	0.20	0.35	234

where there might not be one exact definition for a single metropolitan area. Alternatives would have been to use robustness tests, for example, mining data with a 25-mile radius as well; but again, the method used is so new that it is difficult to determine exact parameters in the code. The mining process itself is also highly resource-consuming. Therefore, we made a restrictive research decision to use 30 miles. A final problem comes from the language used: English-speaking areas would have been overrepresented in the data. Therefore, cities from the United Kingdom were excluded from our analysis. Ireland is included in the analysis, because besides English, it has Irish and Ulster Scots as official languages. The major problem is that in some countries, some words can be spelled the same way as they are in English.

After considering the weaknesses of the model, we considered that the selected approach was the best way to collect the data. Advice was asked also from the peer-users of R from the R-Help mailing list. Considering all of this, it is recognized that perhaps the greater limitation (in terms of data reliability) of the analysis comes from the data collected from Eurostat. Comparing real-time data from SNS to data on economic development that is four years old neglects, for example, the effect of recent growth in e-capital and the economy. However, these shortcomings have been considered and acknowledged.

Spatial Analysis Tools

The analysis is expanded to include the digital divide approach (e.g., Chen, 2013; 2014). The digital divide in Europe is answered with Moran I bivariate statistics, where the two variables are current GDP and predicted GDP. The predicted value is based on e-capital measures. This provided information on cities that perform poorly in both economic and e-capital values. They could be called as "back runners" of the digitalization process which creates digital divide and inequality in the way that cities with low economic development do not possess e-capital either. Instead, cities with high e-capital normally also hold higher levels of economic wealth and other forms of capital.

The first analytical tool applied is standard multivariate regression analysis. We produced four models, in which the number (N) of different keyword tweets are the predictors (independent variables). There are eight predictors, which are the absolute number of innovation, tech, startup, and tolerance tweets, as well as the number of innovation, tech, startup, and tolerance tweets related to population of the metropolitan area. The predicted entity is economic capital, which is represented with four variables: total GDP, GDP per capita, total employment, and employment rate of the metropolitan area. First, R values indicate whether the explanation level that e-capital provides for GDP: The predicted GDP scores of the regression analysis indicate e-capital clusters in Europe (assessed with tweets). Attention is also paid to anomalies. In other words, the analysis provides an insight to potential growth areas that are formed of cities with lower economic activity than a city's e-capital level would suggest. Therefore, the second research question may be answered with interpretations obtainable through residual values.

Spatial analyses were made with open source software GeoDa, which is built to find spatial clusters. Moran I bivariate finds spatial clusters regarding two variables: in this case, the sum variables of e-capital and economic variables. The four nearest neighboring metropolitan areas were included in the spatial models. For example, for Helsinki, the four closest metropolises are from three countries (Tampere and Turku in Finland, Tallinn in Estonia, and Stockholm in Sweden). Similarly, in smaller, and in most cases, more urbanized countries, such as the Netherlands, all four cities included in the analysis were within the same country. We focused on the clusters with statistical significance under 0.05.

Results

Tweet Modeling with Economic Data

Results are discussed in relation to a digital divide (observed differences in the presence of tweets) and e-capital's role as an accelerator of economic growth (urban development). The task is to investigate which metropolitan areas have managed to capitalize their e-capital into economic growth. Alternatively, where areas are performing better than the level of e-capital would suggest, this indicates the presence of other forms of capital that have agglomerated in them. They are likely to grow their economy if only e-capital develops to the same levels as other forms of capital. Areas with low levels of e-capital and lower GDP than expected, in turn, could grow their other forms of capital, i.e., human, social, and cultural capital by "jumping into" the e-capital conversation process. This could succeed possibly with public investments in digitalization, or in human capital, which produces e-capital by increasing the know-how (primarily education level) of the population in question. Similarly, investing in other forms of capital could turn into electronic capital.

The answer to the first research question is that e-capital co-exists with economic success. Regression analysis where GDP per capita was predicted with the proportional and absolute number of tweets related to innovations was highly significant (sig. <0.001) (See Table 2). Still, with the used indicators only 45.0 percent (adjusted R square) of the economic development could be explained. The explanatory power is small, but it indicates a significant (however weak) positive relationship between e-capital and the economy (absolute and relative levels of GDP). It is also important to keep in mind that e-capital does not alone generate wealth. There are numerous studies indicating the strong relationship with human capital (education) and innovation measures that provide more significant interdependence with economic variables. For example, Weckroth and Kemppainen (2016) and Kiuru and Inkinen (2017) have reported these strong correlations in their empirical studies. However, |Whiteley (2000) argued on behalf of the significance of social capital and that its impact is at least on the same level, if not stronger, when compared to human capital indicators. Therefore, we made three more regression analyses to find out if e-capital explains even more of economic development than the 45.0 percent of the first regression model.

The four regression models are summarized in Table 3. The most interesting finding is that Twitter activity concerning the keyword "tech" turned out to be negative in the second model. This would mean that their presence decreases an area's GDP and employment levels in those models. An explanation of this unexpected result could be that the keyword presents a widely used abbreviation of technology. However, tech tweets per population associated positively with absolute GDP suggesting the extensiveness of technology innovations controversial as in earlier literature (Fernández-Olmos and Ramírez-Alesón, 2017). Also the relative number of "innovation" tweets in Model 2 and the relative

Table 3. Constructed four regression models exploring the relationship between tweets and economic/employment statistics. All presented variables are significant in at least one model

Indicator (tweets in Twitter)	Model 1: Relative economic development (t-statistics)	Model 2: Absolute economic development (t-statistics)	Model 3: Relative employment (t-statistics)	Model 4: Absolute employment (t-statistics)
Constant	27.797 (***)	13.684 (***)	71.460 (***)	17.081 (***)
Innovation (absolute number)	1.951	5.891 (***)	0.411	3.045 (*)
Startup (absolute number)	1.007	8.309 (***)	0.272	7.324 (***)
Refugeeswelcome (absolute number)	0.863	5.780 (***)	0.928	7.852 (***)
Tech (absolute number)	−2.012	−5.077 (***)	−0.492	−3.424 (*)
Innovation (related to population)	−0.754	−4.198 (***)	−0.600	−2.614
Refugeeswelcome (related to population)	−0.570	−2.291	0.440	−3.409 (*)
Tech (related to population)	1.744	3.224 (**)	0.936	2.196
R	45.1%	92.7%	25.1%	87.7%
R square	20.3%	86.0%	6.3%	77.0%
Adjusted R square	17.5%	85.5%	2.9%	76.2%
Significance	0.000	0.000	0.063	0.000

* = sig. <0.05
** = sig. <0.01
*** = sig. <0.001

number of "refugeeswelcome" tweets in Model 4 are negatively connected with economic indicators. This should be investigated further in the future studies. Other significant variables in models were positively associated with the (absolute) employment and GDP levels. In models where relative GDP and relative employment were the constant variables, none of the single tweet keywords turned out to be significant. Tweets concerning "startups," "innovation," and "refugeeswelcome" predicted (absolute) employment and (absolute) economic development the best. It is also another example why tolerance (of the area) generates economic growth (e.g., Florida, 2002, 2012).

When Twitter activity is compared to employment figures, the explanation power fell significantly from 45 to 25 percent. In addition, models where relative employment was the constant variable, the significance of the test was slightly non-significant (0.063). The study provided also an analysis of the interrelations between (absolute) GDP and employment contrasted with indicators of e-capital. Based on earlier studies and literature, it is necessary to test both types, absolute and relative, indicators. Absolute levels manifest scale and density—factors that have been widely concluded to create economic growth (e.g., Jacobs, 1969; Fu, 2007; Malizia and Montoyab, 2015; Inkinen and Kaakinen, 2016; Kiuru and Inkinen, 2017). The difference between relative and absolute GDP values is highly significant. R value rose from 45 percent of proportional GDP to 93 percent when predicted absolute GDP. With the absolute amount of economic development and the absolute number of tweets, it is reasonable to conclude that a digitalization process is present in those areas with the high GDP. This takes place in the most urbanized areas.

All in all, regression Model 2 was the best. The combined dataset of absolute innovative Twitter activity and tweets per population represents well the absolute GDP of the studied metropolitan areas. Thus, the result of the first research question advances the interlinkage between social capital and economy (Bourdieu, 1986) and the correlation between e-capital and economic indicators (Mandarano et al., 2010; Merisalo, 2016). It is clear that innovative networks seem to generate economic growth as argued in earlier literature (Robson and Bennett, 2000; Fernández-Olmos and Ramírez-Alesón, 2017). It can be concluded that e-capital related to innovativeness adds to urban (economic) development. The linkages are varied in their extent but they are statistically significant.

Locations of E-Capital and Economic Geography

The second research question concerns the economic geographies of e-capital in Europe. This is answered with predicted GDP scores and residual values from the regression analysis. Metropolitan areas with the highest level of e-capital per capita locate almost exclusively in Western Europe, especially in Benelux countries (See Figure 1). There is also evidence of larger clusters that follow the forms of the so called Blue Banana and Golden Banana. The Blue Banana starts in the British Isles and continues in the continent in the valley of Rhine to Southern Germany and Northern Italy. The Golden Banana, or the Sun Belt, stretches along the coast of the Mediterranean from Valencia to Genoa. These areas are also the most populous and urbanized areas in Europe. Cities with the highest level of absolute e-capital are the largest ones like Paris and Berlin (See Figure 2). However, not every large city experiences high levels of e-capital (such as Munich and Lisbon). In addition, some smaller cities (especially in the Blue Banana cluster) hold rather high levels of absolute e-capital.

In the following, the focus is narrowed to single cities. Antwerp has the most innovation tweets (See Table 4). Utrecht has the most "startup" and "tech" tweets, although the latter had negative association with GDP. Therefore, rankings of sum variables of all the tweets

Figure 1. Economic geography of the relative level of e-capital, i.e., the Twitter activity concerning innovation keywords

Figure 2. Economic geography of the absolute level of e-capital, i.e., the Twitter activity concerning innovation keywords

are presented in the Appendix 1. The result is that tweets with different keywords accumulate mostly in same cities. For, example, besides Utrecht, Paris and Wuppertal also rank in the top 10 in all three categories. Antwerp, Amsterdam, The Hague, and Monchengladbach have constant positions in the top 10.

Another important result concerns the residual values. Their interpretations are not a straightforward task and they may be interpreted in different ways. For example, there are several factors contributing to having higher GDP than the analysis predicted. These locations may be seen as areas that have managed to capitalize their digitalization to the utmost. Areas with smaller GDP than predicted may be seen as potential growth areas. The largest gap between predicted GDP per capita and current tweet activity locate especially in Western and Northern Europe. Absolute GDP could grow in these areas as well as in several Southern Europe cities (See Figures 3 and 4 and the full listing in Appendix 2).

Table 4. The city ranks of the most active tweeting metropolises in Europe according to main keywords

Rank	Innovation	Startup	Tech
1	Antwerpen	Utrecht	Utrecht
2	Utrecht	Wuppertal	Paris
3	The Hague	Salzburg	Wuppertal
4	Paris	Milano	Dublin
5	Wuppertal	Mainz	Amsterdam
6	Bruxelles / Brussel	Darmstadt	The Hague
7	Tilburg	Montpellier	Antwerpen
8	Amsterdam	Paris	Mönchengladbach
9	Rotterdam	Wiesbaden	s-Hertogenbosch
10	Mönchengladbach	Aschaffenburg	Pforzheim

Figure 3. Metropoles with the largest differences in measured tweet activity and relative GDP

Another interpretation is that these urban areas are wealthy in terms of other forms of capital (i.e., human capital, social capital, and cultural capital) that are connected to their economic indicators. This interpretation is consistent with earlier studies that have concluded that e-capital is likely to agglomerate in the same areas as other forms of capital.

Figure 4. Metropoles with the largest differences in measured tweet activity and absolute GDP

Therefore, an examination was conducted regarding where economic development is probably due to other forms of capital. Clustering was made with two variables, GDP per capita and the level of e-capital (predicted GDP per capita in regression analysis). We named the results as COMAs, which are categorized as follows:

(1) High economic performance areas with high e-capital
(2) Low economic performance areas with low e-capital
(3) Low economic performance areas despite high e-capital
(4) High economic performance areas with low e-capital.

The first group, "High economic performance areas with high e-capital" cluster almost exclusively in Benelux countries (See Figure 5). Other well-to-do cities in terms of economy and e-capital are Mainz and Cologne in Germany, Genoa in Italy, and Bergen in Norway. This is an example that when the economy in an area is doing well, (as manifested by high economic and employment figures), that area is also likely to promote digital alternatives and innovations more extensively than those areas struggling with low GDP and high unemployment. This is clearly visible in the maps of Figures 3, 4, and 5, presenting a clear division between the cities of eastern and western Europe. Thus, e-capital analysis also brings forth long historical trajectories in regional economic development. The fourth group consists of cities that are doing well in economic terms despite the lack of e-capital. These cities are Madrid, Berlin, Wien, Hamburg, Nurnberg, and Bayreuth. Based on explanation levels of regression analyses, these cities have growth potentials in terms of e-capital. Another potential growth in GDP per capita can be forecasted for cities belonging to group three, "Low economic performance areas despite high e-capital." There are only two cities in this

Figure 5. A fourfold categorization of European metropolises (COMAs)

group: Perpignan and Amiens in France. Remembering that the economic data on this research was four years older than e-capital measures, these cities may have grown already in economic terms due the high level of e-capital. The fourth group is the most problematic one. This statistically significant group answers to other interpretations of digitalization: the digital divide approach (e.g., Chen and Wellman, 2004; Chen, 2013, 2014). However, if these cities manage to invest in digitalization, their economies could benefit the most due to the current low levels of GDP per capita.

The cities with the most potential to grow their GDP per capita cluster in Germany, Spain, Austria, and France. The most problematic COMA (group 2) covers most of Eastern Europe and some of Southern Europe. These areas are also problematic because English is just one language option and there is no data concerning the Twitter activity in the native languages of these countries. It seems that a number of Eastern European smaller economies have limited participation in Anglo-driven social media use and verifiably in Twitter. Still, there are some well ranking cities in Eastern Europe such as Warsaw, the capital of Poland is significantly rich in e-capital.

Conclusions

The empirical results verified that social capital (Bourdieu, 1986), innovative networks (Miguelez and Moreno, 2013),and "digital social capital" (Mandarano et al., 2010) are associated with "electronic capital" (Merisalo et al., 2013; Merisalo, 2016; Inkinen et al., 2018). This paper has provided one of the first attempts to operationalize the concept of e-capital into new types of empirical measures (beyond those presented by Merisalo, 2016). The applied indicators of the study are related to some extent with metropolitan economic statistics (See Table 3). However, there are significant variations among the studied variables. The absolute number of tweets concerning innovation together with tweets per population had explanation power of 45 percent on GDP per capita. When Twitter activity is compared to the employment figures, the explanation power fell significantly from 45 to 25 percent. E-capital does not affect employment as much as GDP levels. However, it is acknowledged that the nature of the data does not give the opportunity to verify these connections fully as causal: economic growth should also lead to higher e-capital, which again, should lead to further growth. Theoretically we could make an interpretation regarding a "virtuous cycle," but it requires further research to verify the usefulness of Twitter data in approximating the e-capital concept in terms of causal relations.

The study considered the absolute GDP and absolute number of tweets in metropolitan regions motivated by the findings of economic growth in dense neighborhoods in the earlier literature (e.g., Fu, 2007; Malizia and Montoyab, 2015; Inkinen and Kaakinen, 2016; Kiuru and Inkinen, 2017). In fact, it turned out that the best model was obtained with tweets predicting absolute GDP that reached the explanation rate of 92 percent. Regarding spatial scaling and applied regional level, our study illustrates that innovative growth takes place in the most urbanized metropolitan areas as an aggregate, not just the most urban neighborhoods.

In the modeling, the absolute number of "innovation" and "tech" tweets had negative influence on the economy. One reason could be the use of abbreviation of technology.

A second reason could be that a number of persons and organizations, in the field of technology, do not use the social media technologies, particularly Twitter, extensively concerning their profession or innovations. Anyhow, the result indicates a need for further empirical studies. The relative number of "tech" tweets per capita is also related positively with absolute GDP. All this is in line with earlier studies, which have found controversial results of the networks between (SME) firms (in the field of technology) and their productivity and capability creating new innovations (see Ala-Rämi and Inkinen, 2008; Fernández-Olmos and Ramírez-Alesón, 2017).

E-capital in Europe clusters in the most urbanized and populated areas of Europe, also known as the Blue Banana and the Golden Banana. This finding suggests that neighboring cities also have an impact on Twitter activity in European cities. With spatial statistics built to find spatial clusters, we found four new clusters, which we named COMAs. There are some key properties defining them. First, COMAs may be formed by using large metropolitan areas. Second, all COMAs have a neighboring effect. Third, COMAs are useful in predicting the e-capital (tweet measures) and economic success (GDP per capita) in their specific areas across Europe. The main result is that potential growth areas are clustered in Germany, Spain, Austria, and France. Residual values, however, tell only one part of the complex interrelations between technology use and GDP.

Economically high performing cities that have smaller amounts of Twitter activity (in comparison to their peer cities in terms of GDP measures) concerning innovation and technology are Madrid, Berlin, Wien, Hamburg, Nurnberg, and Bayreuth. The second identified COMA (See Figure 5) is the most problematic one. It covers most of Eastern Europe and parts of Southern Europe. Adding another variable into the analysis brought us statistically significant results of cities with low e-capital and low economic indicators. This low-low group is characterized by a lack of other forms of capital as could be concluded also from earlier research (e.g., Merisalo, 2016). This raises a further need to investigate digitalization with broader data. In addition, processes of digitalization and information society development on smaller economic scales are also needed (e.g., Inkinen and Jauhiainen, 2007).

Acknowledgments

The authors thank three anonymous referees for their constructive comments on improving the paper.

Funding

This research was funded by the Helsinki Metropolitan Region Urban Research Program.

ORCID

Tommi Inkinen ⓘ http://orcid.org/0000-0001-6682-043X

References

K. Ala-Rämi and T. Inkinen, "Information Technology, Communication and Networking in Small and Medium Size Software Companies: The Case of Northern Finland," *International Journal of Knowledge Management Studies* 2: 3 (2008) 320–334.

J. Boase, "The Consequences of Personal Networks for Internet Use in Rural Areas," *American Behavioral Scientist* 53: 9 (2010) 1257–1267.

P. Bourdieu, "The Forms of Capital," in I. Szeman, and T. Kaposy, eds, *Cultural Theory* (London: Wiley-Blackwell, 1986).

F. Bougrain and B. Haudeville, "Innovation, Collaboration and SMEs Internal Research Capacities," *Research Policy* 31: 5 (2002) 735–747.

R.S. Burt, *Structural Holes: The Structure of Social Capital Competition* (Cambridge, MA: Harvard University Press, 1992).

W. Chen, "The Implications of Social Capital for the Digital Divides in America," *The Information Society. An International Journal* 29: 1 (2013) 13–25.

W. Chen, J. Boase, and B. Wellman, "The Global Villagers: Comparing Internet Users and Uses around the World," in B. Wellman, and C. Haythornthwaite, eds, *The Internet in Everyday Life* (Oxford: Blackwell, 2002).

W. Chen and B. Wellman, "The Global Digital Divide: Within and Between Countries," *IT & Society* 1: 7 (2004) 18–25.

T.N. Clark, "Urban Amenities: Lakes, Opera, and Juice Bars: Do They Drive Development?" in T.N. Clark, ed., *The City as an Entertainment Machine* (Bingley: Emerald, 2003) 103–140.

P. Cooke, "Regional Innovation Systems: Competitive Regulation in the New Europe," *Geoforum* 23: 3 (1992) 365–382.

P. Cooke, "Complex Spaces: Global Innovation Networks & Territorial Innovation Systems in Information & Communication Technologies," *Journal of Open Innovation: Technology, Market, and Complexity* 3: 9 (2017) 1–23.

J. Cranshaw, R. Schwartz, J. Hong, and N. Sadeh, "The Livehoods Project: Utilizing Social Media to Understand the Dynamics of a City," paper presented at the Sixth International AAAI Conference on Weblogs and Social Media (Dublin, June 4–7, 2012).

CyBranding Ltd., *Hashtagify.me* (London, 2016) Accessed January 31, 2019.

P. DiMaggio, E. Hargittai, W.R. Neuman, and J.P. Robinson, "Social Implications of the Internet," *Annual Review of Sociology* 27: 2 (2001) 307–336.

S. Dutta, B. Lanvin, and S. Wunsch-Vincent, *Global Innovation Index: Innovation Feeding the World* (Geneva: Cornell University, INSEAD, and the World Intellectual Property Organization, 2017).

B.H. Erickson, "Culture, Class, and Connections," *American Journal of Sociology* 102: 1 (1996) 217–251.

European Commission, *What Is a City? – Spatial Units* (Brussels, 2016a) <http://ec.europa.eu/eurostat/web/cities/spatial-units> Accessed January 31, 2019.

European Commission, *Metropolitan Regions* (Brussels, 2016b) <http://ec.europa.eu/eurostat/web/metropolitan-regions> Accessed January 31, 2019.

M. Fernández-Olmos and M. Ramírez-Alesón, "How Internal and External Factors Influence the Dynamics of SME Technology Collaboration Networks Over Time," *Technovation* 64 (2017) 16–27.

R. Florida, *The Rise of the Creative Class* (New York: Basic Books, 2002).

R. Florida, *The Rise of the Creative Class: Revisited* (New York: Basic Books, 2012).

M. Franco and H. Haase, "Interfirm Alliances: A Taxonomy for SMEs," *Long Range Planning* 48: 3 (2015) 168–181.

S. Fu, "Smart Café Cities: Testing Human Capital Externalities in the Boston Metropolitan Area," *Journal of Urban Economics* 61: 1 (2007) 86–111.

Github, *Geoffjentry* (San Francisco, 2016) <https://github.com/geoffjentry> Accessed January 31, 2019.

E.L. Glaeser, R. La Porta, F. Lopez-de-Silanes, and A. Shleifer, "Do Institutions Cause Growth?" *Journal of Economic Growth* 9: 3 (2004) 271–303.

M.S. Granovetter, "The Strength of Weak Ties," *American Journal of Sociology* 78: 13 (1973) 1360–1380.

R.E. Hall, "E-Capital: The Link between the Stock Market and the Labor Market in the 1990s," *Brookings Papers on Economic Activity* 2000: 2 (2000) 73–102.

S. Hasan, X. Zhan, and V. Ukkusuri, "Understanding Urban Human Activity and Mobility Patterns Using Large-Scale Location-Based Data from Online Social Media," paper presented at UrbComp of the 2nd ACM SIGKDD International Workshop on Urban Computing (Chicago, August 11–14, 2013).

J.F. Helliwell and R. Putnam, "Economic Growth and Social Capital in Italy," *Eastern Economic Journal* 21: 3 (1995) 295–307.

T. Hiippala, A. Hausmann, H.T.O. Tenkanen, and T.K. Toivonen, "Exploring the Linguistic Landscape of Geotagged Social Media Content in Urban Environments," *Digital Scholarship in the Humanities: DSH* (2018).

T. Inkinen, and J.S. Jauhiainen, "Public Authorities and the Local Information Society," in A.-V. Anttiroiko, and M. Mälkiä, eds, *Encyclopedia of Digital Government* (Hershey: Idea Group Reference, 2006).

T. Inkinen, and I. Kaakinen, "Economic Geography of Knowledge Intensive Technology Clusters: Lessons from the Helsinki Metropolitan Area," *Journal of Urban Technology* 23: 1 (2016) 95–114.

T. Inkinen, M. Merisalo, and T. Makkonen, "Variations in the Adoption and Willingness to Use E-Services in Three Differentiated Urban Areas," *European Planning Studies* 26: 5 (2018) 950–968.

S. Iyer, M. Kitson, and B. Toh, "Social Capital, Economic Growth and Regional Development," *Regional Studies* 39: 8 (2005) 1015–1040.

J. Jacobs, *The Economies of Cities* (New York: Random House, 1969).

J. Kiuru and T. Inkinen, "Predicting Innovative Growth and Demand with Proximate Human Capital: A Case Study of the Helsinki Metropolitan Area," *Cities* 64 (2017) 9–17.

S. Knack and P. Keefer, "Does Social Capital Have an Economic Payoff? A Cross-Country Investigation," *The Quarterly Journal of Economics* 112: 4 (1997) 1251–1288.

R. LaRose, J. L. Gregg, S. Strover, J. Straubhaar, and N. Inagaki, *Closing the Rural Broadband Gap (Final Technical Report US Department of Agriculture CREES Program, Grant Number 2004–35401-14985* (East Lansing: Department of Telecommunication, Information Studies, and Media, Michigan State University, 2008).

E. Malizia and Y. Motoyama, "The Economic Development–Vibrant Center Connection: Tracking High-Growth Firms in the DC Region," *The Professional Geographer* 68: 3 (2015) 349–355.

L. Mandarano, M. Mahbubur, and C. Steins "Building Social Capital in the Digital Age of Civic Engagement," *Journal of Planning Literature* 25: 2 (2010) 123–135.

M. Merisalo, *Electronic Capital: Economic and Social Geographies of Digitalization* (Helsinki: University of Helsinki, Faculty of Science, Department of Geosciences and Geography, 2016).

M. Merisalo, T. Makkonen, and T. Inkinen, "Creative and Knowledge-Intensive Teleworkers' Relation to E-Capital in the Helsinki Metropolitan Area," *International Journal of Knowledge-Based Development* 4: 3 (2013) 204–221.

E. Miguelez and R. Moreno, "Research Networks and Inventors' Mobility as Drivers of Innovation Evidence from Europe," *Regional Studies* 47: 10 (2013) 1668–1685.

A. Niebuhr, "Migration and Innovation: Does Cultural Diversity Matter for Regional R&D Activity?" *Papers in Regional Science* 89: 3 (2010) 563–585.

M.J. Nieto and L. Santamaría, "The Importance of Diverse Collaborative Networks for the Novelty of Product Innovation," *Technovation* 27: 6 (2007) 367–377.

J.M. Podolny, "A Status-Based Model of Market Competition," *American Journal of Sociology* 98: 4 (1993) 829–872.

M. Porter, "Location, Competition, and Economic Development: Local Clusters in a Global Economy," *Economic Development Quarterly* 14: 1 (2000) 15–34.

R.D. Putnam, "Bowling Alone: America's Declining Social Capital," *Journal of Democracy* 6: 1 (1995) 65–78.

R.D. Putnam, "Bowling Alone: America's Declining Social Capital," in L. Crothers and C. Lockhart, eds, *Culture and Politics* (New York: Springer, 2000).

H. Qian, "Diversity Versus Tolerance: The Social Drivers of Innovation and Entrepreneurship in US Cities," *Urban Studies* 50: 13 (2013) 2718–2735.

P.J.A. Robson and R.J. Bennet, "SME Growth: The Relationship with Business Advice and External Collaboration," *Small Business Economics* 15: 3 (2000) 193–208.

U. Rossi, *Cities in Global Capitalism* (London: Polity Press, 2016).

L. Song, "Social Capital and Health," in W.C. Cockerham, ed., *Medical Sociology on the Move* (New York: Springer, 2013).

K.-H. Tsai, "Collaborative Networks and Product Innovation Performance: Toward a Contingency Perspective," *Research Policy* 38: 5 (2009) 765–778.

M. Weckroth and T. Kemppainen, "Human Capital, Cultural Values and Economic Performance in European Regions," *Regional Studies, Regional Science* 3: 1 (2016) 239–257.

P. F. Whiteley, "Economic Growth and Social Capital," *Political Studies* 48: 3 (2000) 443–466.

M. Woolcock, "The Place of Social Capital in Understanding Social and Economic Outcomes," *Canadian Journal of Policy Research* 2: 1 (2001) 11–17.

N. Zillien and E. Hargittai, "Digital Distinction: Status-Specific Types of Internet Usage," *Social Science Quarterly* 90 (2009) 274–291.

D.B. Zwiefelhofer, *Find Latitude and Longitude* (Milwaukee, 2016) <http://www.findlatitudeandlongitude.com> Accessed January 31, 2019.

Appendix 1. Listing of the most active metropolitan areas of Europe in terms of tweets (constructed e-capital measure)

Rank	Metro area	E-capital	Rank	Metro area	E-capital
1	Utrecht	5.7	118	Besançon	−0.4
2	Antwerpen	5.1	119	Tampere	−0.4
3	Bruxelles / Brussel	4.8	120	Clermont-Ferrand	−0.4
4	Paris	4.4	121	Rouen	−0.4
5	Wuppertal	3.6	122	Murcia - Cartagena	−0.4
6	Den Hague	3.0	123	Wetzlar	−0.4
7	Milano	2.8	124	Gießen	−0.4
8	Amsterdam	2.4	125	Odense	−0.5
9	Lyon	1.9	126	Toulon	−0.5
10	Rotterdam	1.8	127	Enschede	−0.5
11	Berlin	1.8	128	Århus	−0.5
12	Toulouse	1.7	129	Dresden	−0.5
13	Stockholm	1.7	130	Konstanz	−0.5
14	Darmstadt	1.7	131	Kiel	−0.5
15	Dublin	1.6	132	Palma de Mallorca	−0.5
16	Mainz	1.6	133	Bayreuth	−0.5
17	Madrid	1.6	134	Ljubljana	−0.5
18	Prato	1.5	135	Tours	−0.5
19	Barcelona	1.5	136	Paderborn	−0.5
20	Wiesbaden	1.4	137	Koblenz	−0.5
21	Montpellier	1.4	138	Oviedo - Gijón	−0.5
22	Pforzheim	1.4	139	Vilnius	−0.5
23	Mönchengladbach	1.4	140	Messina	−0.5
24	Reutlingen	1.4	141	Kraków	−0.5

(Continued)

Continued.

Rank	Metro area	E-capital	Rank	Metro area	E-capital
25	Salzburg	1.4	142	Münster	−0.5
26	s-Hertogenbosch	1.3	143	Santander	−0.5
27	Aschaffenburg	1.2	144	Saint-Etienne	−0.5
28	Roma	1.2	145	Zagreb	−0.5
29	Köln	1.1	146	Sevilla	−0.5
30	Nantes	1.1	147	Taranto	−0.5
31	Frankfurt am Main	1.0	148	Amiens	−0.5
32	Tilburg	0.9	149	Alicante/Alacant - Elche/Elx	−0.5
33	Bonn	0.8	150	Iasi	−0.5
34	Liège	0.8	151	Bremen	−0.5
35	Nimes	0.7	152	Nürnberg	−0.5
36	Helsinki	0.7	153	Pamplona/Iruña	−0.5
37	Hamburg	0.7	154	Bielefeld	−0.5
38	Malmö	0.6	155	Marburg	−0.5
39	Nice	0.6	156	Córdoba	−0.5
40	Düsseldorf	0.6	157	Bergen	−0.5
41	Bordeaux	0.5	158	Linz	−0.5
42	Eindhoven	0.5	159	Le Mans	−0.5
43	Wien	0.5	160	Brest	−0.5
44	Rennes	0.5	161	Wroclaw	−0.5
45	København	0.4	162	Limoges	−0.5
46	Heilbronn	0.4	163	A Coruña	−0.5
47	Offenburg	0.4	164	Aalborg	−0.5
48	Lille - Dunkerque - Valenciennes	0.3	165	Pau	−0.5
49	Padova	0.3	166	Sofia	−0.5
50	Marseille	0.3	167	Thessaloniki	−0.5
51	Venezia	0.3	168	Dijon	−0.5
52	Modena	0.2	169	Poznan	−0.5
53	Arnhem - Nijmegen	0.2	170	Kassel	−0.5
54	Hildesheim	0.2	171	Bari	−0.5
55	Bologna	0.2	172	Innsbruck	−0.5
56	München	0.2	173	Görlitz	−0.5
57	Rosenheim	0.1	174	Braunschweig-Salzgitter-Wolfsburg	−0.5
58	Charleroi	0.1	175	Brescia	−0.5
59	Parma	0.1	176	Vigo	−0.5
60	Warszawa	0.1	177	Brno	−0.5
61	Breda	0.0	178	Plzen	−0.5
62	Cork	0.0	179	Poitiers	−0.6
63	Zaragoza	0.0	180	Siegen	−0.6
64	Valencia	0.0	181	Saarbrücken	−0.6
65	Stuttgart	0.0	182	Cádiz - Algeciras	−0.6
66	Genova	−0.1	183	Valletta	−0.6
67	Torino	−0.1	184	Kaiserslautern	−0.6
68	Aachen	−0.1	185	Lübeck	−0.6
69	Basel	−0.1	186	Ulm	−0.6
70	Ruhrgebiet	−0.1	187	Osnabrück	−0.6
71	Tallinn	−0.1	188	Göttingen	−0.6
72	Valladolid	−0.1	189	Ingolstadt	−0.6
73	Strasbourg	−0.1	190	Gdansk	−0.6
74	Hannover	−0.1	191	Plauen	−0.6
75	Heerlen	−0.1	192	Maribor	−0.6
76	Firenze	−0.1	193	Erfurt	−0.6
77	Napoli	−0.2	194	Lódz	−0.6
78	Grenoble	−0.2	195	Bielsko-Biala	−0.6
79	Uppsala	−0.2	196	Bremerhaven	−0.6
80	Gent	−0.2	197	Budapest	−0.6
81	Reims	−0.2	198	Constanta	−0.6
82	Oslo	−0.2	199	Pécs	−0.6
83	Groningen	−0.2	200	Katowice	−0.6
84	Bilbao	−0.3	201	Oldenburg (Oldenburg)	−0.6
85	Caen	−0.3	202	Ostrava	−0.6

(Continued)

Continued.

Rank	Metro area	E-capital	Rank	Metro area	E-capital
86	Augsburg	−0.3	203	Kaunas	−0.6
87	Mulhouse	−0.3	204	Split	−0.6
88	Mannheim-Ludwigshafen	−0.3	205	Galati	−0.6
89	Angers	−0.3	206	Schweinfurt	−0.6
90	Bratislava	−0.3	207	Szczecin	−0.6
91	Catania	−0.3	208	Brasov	−0.6
92	Nancy	−0.3	209	Rostock	−0.6
93	Riga	−0.3	210	Rzeszów	−0.6
94	Zwickau	−0.3	211	Magdeburg	−0.6
95	Göteborg	−0.3	212	Kosice	−0.6
96	Málaga - Marbella	−0.3	213	Tarnów	−0.6
97	Iserlohn	−0.3	214	Bydgoszcz - Torún	−0.6
98	Santa Cruz de Tenerife	−0.3	215	Plovdiv	−0.6
99	Freiburg im Breisgau	−0.4	216	Flensburg	−0.6
100	Perpignan	−0.4	217	Varna	−0.6
101	Heidelberg	−0.4	218	Würzburg	−0.6
102	Verona	−0.4	219	Debrecen	−0.6
103	Turku	−0.4	220	Lublin	−0.6
104	Cagliari	−0.4	221	Opole	−0.6
105	Karlsruhe	−0.4	222	Radom	−0.6
106	Donostia-San Sebastián	−0.4	223	Székesfehérvár	−0.6
107	Orléans	−0.4	224	Czestochowa	−0.6
108	Halle an der Saale	−0.4	225	Kielce	−0.6
109	Leipzig	−0.4	226	Miskolc	−0.6
110	Cluj-Napoca	−0.4	227	Craiova	−0.6
111	Palermo	−0.4	228	Bialystok	−0.6
112	Graz	−0.4	229	Schwerin	−0.6
113	Avignon	−0.4	230	Neubrandenburg	−0.6
114	Timisoara	−0.4	231	Bucuresti	−0.6
115	Las Palmas	−0.4	232	Athina	−0.6
116	Granada	−0.4	233	Vitoria/Gasteiz	−0.7
117	Regensburg	−0.4	234	Praha	−0.7

Appendix 2. Metropolitan areas having higher economic performance than their e-capital level would suggest

Rank	Metro area	E-capital gap	Rank	Metro area	E-capital gap
1	Oslo	4.8	118	Erfurt	0.1
2	Bergen	3.4	119	Rennes	0.1
3	Groningen	3.3	120	Amiens	0.1
4	Ingolstadt	2.3	121	Besançon	0.1
5	Stockholm	2.2	122	Venezia	0.0
6	München	2.1	123	Leipzig	0.0
7	København	2.0	124	Brest	0.0
8	Cork	1.8	125	Nancy	0.0
9	Düsseldorf	1.5	126	Kaiserslautern	0.0
10	Göteborg	1.5	127	Mulhouse	0.0
11	Stuttgart	1.4	128	Limoges	0.0
12	Linz	1.4	129	Athina	0.0
13	Innsbruck	1.4	130	Angers	−0.1
14	Helsinki	1.4	131	Ljubljana	−0.1
15	Mannheim-Ludwigshafen	1.4	132	Tilburg	−0.1
16	Karlsruhe	1.4	133	Paris	−0.1
17	Münster	1.4	134	Dresden	−0.1
18	Heilbronn	1.3	135	Palma de Mallorca	−0.1

(Continued)

Continued.

Rank	Metro area	E-capital gap	Rank	Metro area	E-capital gap
19	Uppsala	1.3	136	Toulouse	−0.1
20	Regensburg	1.3	137	Schwerin	−0.1
21	Århus	1.2	138	Reutlingen	−0.1
22	Nürnberg	1.2	139	Toulon	−0.1
23	Braunschweig-Salzgitter-Wolfsburg	1.2	140	Praha	−0.1
24	Frankfurt am Main	1.2	141	Nantes	−0.2
25	Salzburg	1.2	142	Lille - Dunkerque - Valenciennes	−0.2
26	Aalborg	1.1	143	Zwickau	−0.2
27	Schweinfurt	1.1	144	Neubrandenburg	−0.2
28	Ulm	1.1	145	Görlitz	−0.2
29	Graz	1.1	146	Darmstadt	−0.3
30	Koblenz	1.1	147	Zaragoza	−0.3
31	Eindhoven	1.0	148	Perpignan	−0.3
32	Gent	1.0	149	Cagliari	−0.3
33	Wien	1.0	150	Plauen	−0.3
34	Breda	0.9	151	A Coruña	−0.4
35	Dublin	0.9	152	s' Gravenhage	−0.4
36	Hannover	0.9	153	Santander	−0.4
37	Tampere	0.8	154	Pforzheim	−0.4
38	Vitoria/Gasteiz	0.8	155	Hildesheim	−0.4
39	Odense	0.8	156	Valladolid	−0.4
40	Hamburg	0.8	157	Oviedo - Gijón	−0.4
41	Bielefeld	0.8	158	Bari	−0.5
42	Saarbrücken	0.8	159	Charleroi	−0.5
43	Bologna	0.8	160	Mönchengladbach	−0.5
44	Kassel	0.7	161	Liège	−0.5
45	Würzburg	0.7	162	Madrid	−0.5
46	Bremen	0.7	163	Vigo	−0.5
47	Turku	0.7	164	Santa Cruz de Tenerife	−0.6
48	Firenze	0.6	165	Sevilla	−0.6
49	Siegen	0.6	166	Las Palmas	−0.6
50	Dijon	0.6	167	Valencia	−0.6
51	Bayreuth	0.6	168	Murcia - Cartagena	−0.6
52	Augsburg	0.6	169	Valletta	−0.6
53	Paderborn	0.6	170	Alicante/Alacant - Elche/Elx	−0.7
54	Osnabrück	0.5	171	Taranto	−0.7
55	Reims	0.5	172	Montpellier	−0.7
56	Malmö	0.5	173	Palermo	−0.7
57	Amsterdam	0.5	174	Warszawa	−0.7
58	Bratislava	0.5	175	Messina	−0.7
59	Heidelberg	0.5	176	Tallinn	−0.7
60	Gießen	0.5	177	Cádiz - Algeciras	−0.8
61	Iserlohn	0.5	178	Budapest	−0.8
62	Bonn	0.5	179	Bucuresti	−0.8
63	Brescia	0.5	180	Prato	−0.8
64	Parma	0.5	181	Córdoba	−0.8
65	Köln	0.5	182	Napoli	−0.8
66	Freiburg im Breisgau	0.5	183	Vilnius	−0.8
67	Wetzlar	0.5	184	Berlin	−0.8
68	Orléans	0.4	185	Poznan	−0.8
69	Marburg	0.4	186	Granada	−0.8
70	Offenburg	0.4	187	Bruxelles / Brussel	−0.8
71	Verona	0.4	188	Catania	−0.8
72	Enschede	0.4	189	Málaga - Marbella	−0.8
73	Rouen	0.4	190	Zagreb	−0.9
74	Genova	0.4	191	Nimes	−0.9
75	Clermont-Ferrand	0.4	192	Barcelona	−0.9
76	Donostia-San Sebastián	0.4	193	Thessaloniki	−0.9
77	Wiesbaden	0.4	194	Brno	−0.9
78	Strasbourg	0.4	195	Maribor	−0.9
79	Pau	0.4	196	Plzen	−1.0

(Continued)

Continued.

Rank	Metro area	E-capital gap	Rank	Metro area	E-capital gap
80	Arnhem - Nijmegen	0.4	197	Ostrava	−1.0
81	Göttingen	0.4	198	Riga	−1.0
82	Heerlen	0.3	199	Wroclaw	−1.1
83	Marseille	0.3	200	Antwerpen	−1.2
84	Oldenburg (Oldenburg)	0.3	201	Gdansk	−1.2
85	Grenoble	0.3	202	Katowice	−1.2
86	Modena	0.3	203	Kaunas	−1.2
87	Lyon	0.3	204	Kraków	−1.2
88	Lübeck	0.3	205	Lódz	−1.2
89	Kiel	0.3	206	Bydgoszcz - Torún	−1.2
90	Rotterdam	0.3	207	Kosice	−1.3
91	Konstanz	0.3	208	Sofia	−1.3
92	Aachen	0.3	209	Szczecin	−1.3
93	Caen	0.2	210	Utrecht	−1.3
94	Ruhrgebiet	0.2	211	Bielsko-Biala	−1.3
95	Basel	0.2	212	Székesfehérvár	−1.4
96	Torino	0.2	213	Opole	−1.4
97	Pamplona/Iruña	0.2	214	Lublin	−1.4
98	Le Mans	0.2	215	Rzeszów	−1.4
99	s-Hertogenbosch	0.2	216	Bialystok	−1.4
100	Tours	0.2	217	Czestochowa	−1.4
101	Roma	0.2	218	Brasov	−1.4
102	Avignon	0.2	219	Kielce	−1.5
103	Mainz	0.2	220	Constanta	−1.5
104	Poitiers	0.2	221	Split	−1.5
105	Rostock	0.2	222	Timisoara	−1.5
106	Padova	0.2	223	Wuppertal	−1.5
107	Flensburg	0.2	224	Debrecen	−1.5
108	Magdeburg	0.1	225	Cluj-Napoca	−1.5
109	Milano	0.1	226	Radom	−1.5
110	Rosenheim	0.1	227	Tarnów	−1.6
111	Saint-Etienne	0.1	228	Pécs	−1.6
112	Bilbao	0.1	229	Miskolc	−1.6
113	Nice	0.1	230	Varna	−1.7
114	Halle an der Saale	0.1	231	Craiova	−1.7
115	Bordeaux	0.1	232	Galati	−1.8
116	Bremerhaven	0.1	233	Iasi	−1.8
117	Aschaffenburg	0.1	234	Plovdiv	−1.8

How to Overcome the Dichotomous Nature of Smart City Research: Proposed Methodology and Results of a Pilot Study

Luca Mora, Mark Deakin, Alasdair Reid, and Margarita Angelidou

ABSTRACT

Overcoming the dichotomous nature of smart city research is fundamental to providing cities with a clear understanding of how smart city development should be approached. This paper introduces a research methodology for conducting the multiple-case study analyses necessary to meet this challenge. After presenting the methodology, we test the practical feasibility, effectiveness, and logistics of such a methodology by examining the activities that Vienna has implemented in building its smart city development strategy. The results of this pilot study show how the application of the proposed methodology can help smart city researchers codify the knowledge produced from multiple smart city experiences, using a common protocol. This in turn allows them to: (1) coordinate efforts when investigating the strategic principles that drive smart city development and test the divergent hypotheses emerging from the scientific literature; (2) share the results of this investigation and hypothesis testing by conducting extensive cross-case analyses among multiple studies able to capture the generic qualities of the findings; (3) gain consensus on the way to think about, conceptualize, and standardize the analysis of smart city developments; and (4) develop innovative monitoring and evaluation systems for smart city development strategies by reflecting upon the lessons learned from current practices.

Introduction

Mora et al. (2017; 2018a; 2018b) and Komninos and Mora (2018) reveal the presence of a deeply rooted division in the underlying structure of smart city research and show that such a division surfaces as a set of dichotomies that question how smart city development[1] should be approached. Their research demonstrates that these dichotomies present themselves as divergent hypotheses on what strategic principles need to be considered when designing and implementing the actions required to support the development of smart cities. It also suggests that this division generates uncertainty over the way ahead and creates a knowledge gap that needs to be closed in order for smart cities to move as a set of groundbreaking urban innovations to demonstrate the "transformative and disruptive role technology [has] in solving urban issues" and supporting urban sustainability (March, 2016: 1694).

This paper considers multiple-case-study research with a deductive approach as one of the most suitable methods for testing the divergent strategic principles for smart city development that each dichotomy stands for. The ambiguity surrounding smart city research demonstrates that a critical synthesis of the literature produced to date is missing, and the empirical knowledge needed to close the gap that currently exists between theory and practice has yet to be generated. Multiple-case study analyses that are able to investigate the smart city phenomenon under different conditions are required to bring about such knowledge and delineate what strategic principles drive smart city development. The use of a common research design can help smart city researchers coordinate their efforts towards the achievement of this common objective.[2]

Following such a line of reasoning, this paper introduces a research methodology that can be deployed for conducting large-scale multiple-case study analyses of smart city development strategies and acquiring the scientific knowledge necessary to overcome the dichotomies at the heart of smart city research. In addition, it reports on the results of a pilot study that is instrumental in examining the practical feasibility, effectiveness, and logistics of the proposed methodology. During the pilot study, the smart city development strategy implemented by the city of Vienna was examined.

The paper is structured in four sections. The first provides an extensive discussion of the four dichotomies that Mora et al. (2017; 2018a; 2018b) have identified. This discussion is based on the review of the most recent literature on smart cities and offers an understanding of the division within smart city research. The design of the methodology proposed for testing the divergent strategic principles that each dichotomy puts forward is presented in the second section, while its functioning is examined in the third, which reports on the results of the pilot study and the authors' experience of conducting it. Structuring the paper in this fashion makes it possible to show how the proposed methodology can be deployed and how its components work together, ensuring its practical feasibility and effectiveness. The logistics of the pilot study also allow the singular nature of this case study to be addressed by: (1) evaluating the sustainability of the research process and its practicability; (2) providing recommendations that are based on direct experience; and (3) detecting practical issues and limitations that may affect a larger sample of case study analyses and provide possible solutions. These three aspects are discussed in the last and conclusive section of the paper, which addresses the lessons learned from the pilot study.

The Dichotomous Nature of Smart City Research

Transforming urban areas into smart cities is an ambition that local and regional governments are trying to realize by developing strategies that make it possible to tackle urban sustainability by means of ICT solutions. Cases of smart city development strategies can be found in communities all over the world, and their developments have been captured by an increasing number of studies. Some of these studies and the smart city cases they examine are listed in Table 1.

However, despite a growing interest in smart cities and almost three decades of literature analyzing their development, a clear explanation of what needs to be done in order for urban environments to succeed in designing and implementing strategies for supporting smart city transformations is still missing. In the literature currently available, different development paths can be identified (Mora et al., 2017; 2018a; 2018b; Komninos and

Table 1. Some of the smart city developments under investigation

Reference	Smart city cases
Aina 2017	Yanbu Industrial City; Jubail Industrial City; King Abdullah Economic City; Riyadh; Jeddah; Dammam; Makkah; Madinah **(Asia)**
Alawadhi et al. 2012	Mexico City, Philadelphia, Quebec City, Seattle **(North America)**
Anderson et al. 2012	Cape Town **(Africa)**; Dongtan, Gujarat International Financial Tech-City, Jubail, Lavasa, Masdar, Shanghai, Shenyang, Songdo, Suwon, Taoyuan, Tianjin, Urumqi, Wuxi **(Asia)**; Ballarat, Gold Coast City, Ipswich **(Australia)**; Amsterdam, Besançon, Birmingham, Bottrop, Copenhagen, Dublin, Eindhoven, Gdansk, Issy-les-moulineaux, Kalundborg, Lyon, Malaga, Malmö, PlanIT Valley, Rotterdam, Sopron, Tallinn, Trikala, Trondheim **(Europe)**; Bristol, Chattanooga, Cleveland, Dakota County, Dublin, Moncton, Ottawa, Quebec City, Toronto, Windsor-Essex, Winnipeg **(North America)**; Curitiba, Pedra Branca, Porto Alegre, Recife **(South America)**
Angelidou 2014	Singapore, Songdo IBD **(Asia)**; Amsterdam, Barcelona, Malta, Thessaloniki **(Europe)**; New York City **(North America)**; Rio de Janeiro **(South America)**
Angelidou 2017	Konza **(Africa)**; Cyberjaya, Singapore, King Abdullah Economic City, Masdar, Skolkovo, Songdo **(Asia)**; Amsterdam, Barcelona, London, PlanIT Valley, Stockholm **(Europe)**; Chicago, New York **(North America)**; Rio de Janeiro **(South America)**
ARUP 2013	Hong Kong **(Asia)**; Barcelona, Stockholm **(Europe)**; Boston, Chicago **(North America)**; Rio de Janeiro **(South America)**
Bakici et al. 2013	Barcelona **(Europe)**
Bolici and Mora 2015	Amsterdam, Barcelona **(Europe)**
Cardullo and Kitchin 2017	Dublin **(Europe)**
Cisco Systems 2012	Busan, Singapore **(Asia)**; Barcelona, Oulu, Rivas-Vaciamadrid, Stockholm **(Europe)**; Boston, San Francisco, South Bend **(North America)**; Rio de Janeiro **(South America)**
Coletta et al. 2017	Dublin **(Europe)**
Cowley et al. 2017	Bristol, Glasgow, London, Manchester, Milton Keynes, Peterborough **(Europe)**
Cugurullo 2013	Masdar **(Asia)**
Dameri 2014	Amsterdam, Genoa **(Europe)**
Datta 2015	Dholera **(Asia)**
Ferrer 2017	Barcelona **(Europe)**
Fietkiewicz and Stock 2015	Kyoto, Osaka, Tokyo, Yokohama **(Asia)**
Gupta and Hall 2017	Agartala; Agra; Ahmedabad; Ajmer; Aligarh; Allahabad; Amravati; Amritsar; Aurangabad; Bareilly; Belagavi; Belgaum; Bhagalpur; Bhopal; Bhubaneswar; Bidhannagar; Biharsharif; Bilaspur; Chandigarh; Chennai; Coimbatore; Dahod; Davangere; Dharamshala; Dindigul; Erode; Faridabad; Gandhinagar; Ghaziabad; Guwahati; Gwalior; Haldia; Hubballi-Dharwad; Imphal; Indore; Jabalpur; Jaipur; Jalandhar; Jhansi; Kakinada; Kalyan-Dombivali; Kanpur; Karnal; Kochi; Kohima; Kota; Lucknow; Ludhiana; Madurai; Mangaluru; Moradabad; Nagpur; Namchi; Nashik; NDMC; New Town Kolkata; Oulgaret; Panaji; Portblair; Pune; Raipur; Rajkot; Rampur; Ranchi; Rourkela; Sagar; Saharanpur; Salem; Shillong; Shivamogga; Solapur; Surat; Thane; Thanjavur; Thoothukudi; Tiruchirappalli; Tirupati; Tiruppur; Tumakaru; Udaipur; Ujjain; Vadodara; Varanasi; Vellore; Visakhapatnam; Warangal **(Asia)**
Komninos 2011	Hong Kong **(Asia)**; Amsterdam, Milton Keynes **(Europe)**
Lee et al. 2014	Seoul **(Asia)**; San Francisco **(North America)**
Leydesdorff and Deakin (2011)	Edinburgh **(Europe)**; Montreal **(North America)**
Mora and Bolici 2016	Barcelona **(Europe)**
Mora and Bolici 2017	Amsterdam **(Europe)**
Mora et al. 2018c	Amsterdan, Barcelona, Helsinki, Vienna **(Europe)**
Sauer 2012	Amsterdam **(Europe)**
Shwayri 2013	Songdo **(Asia)**
Vanolo 2014	Bari, Bologna, Turin, Genoa, Milan, Naples **(Europe)**
Zygiaris 2013	Amsterdam, Barcelona, Edinburgh **(Europe)**

Mora, 2018), whose presence generates uncertainty on how to approach smart city development. This is because these paths suggest strategic principles that are divergent in nature, making it difficult to establish whether smart city development should be based on a: (1) technology-led or holistic strategy; (2) double or quadruple-helix model of collaboration; (3) top-down or bottom-up approach; (4) mono-dimensional or integrated intervention logic. The questions arising from each dichotomy mark a knowledge gap that current research on smart cities is unable to close, and which can be demonstrated by reviewing the literature produced to date (see Table 2).

When trying to define whether a successful smart city development strategy is technology-led or holistic, smart city researchers provide two different answers (Niaros, 2016; Mora et al., 2017). On the one hand, according to the literature produced by ICT companies such as IBM, Cisco Systems, Siemens, ABB, Hitachi, and Fujitsu, smart city development is driven only by information and communication technologies. On the other hand, a large body of literature suggests this vision is inadequate to support smart city development because it conceives smart cities as technological objects rather than complex socio-technical systems in which technological development needs to be aligned with human, social, cultural, economic, and environmental factors.

In addition, as pointed out by Soderstrom et al. (2014), McNeill (2016) and Paroutis et al. (2014), ICT companies also suggest smart city development strategies require a narrow collaborative model in which the interaction is only between service providers selling their smart city solutions to local governments. However, a large number of researchers consider the double-helix structure of this collaborative model unable to provide the intellectual capital that is necessary to drive smart city development. Their research calls for a much more open and inclusive collaborative ecosystem based on a quadruple-helix structure[3] where all the city stakeholders representing research, industry, and government are involved, along with civil society organizations and citizens.[4]

This division also surfaces in relation to the third question: is the most suitable approach for developing smart cities top-down or bottom-up? In the first case, the city government defines both a long-term vision and a strategic framework for supporting smart city development. Whereas the bottom-up approach is deregulated, based on self-organization, and founded on grassroots movements. In addition, it puts civil society in the driver's seat, suggesting the direct involvement of the people in the development of ICT-driven solutions for urban sustainability and their integration in the urban environment is what determines whether strategies are successful or not. The researchers championing the bottom-up approach also highlight the importance of a radical shift from top-down urban innovation processes and movement towards an open and bottom-up process of urban innovation.

The last dichotomy concerns the intervention logic to be considered when implementing smart city development strategies. According to Manville et al. (2014), smart city development requires an integrated and multi-dimensional approach and a successful smart city development strategy covers a large number of policy domains. This assumption is in line with the assessment system that Vienna University of Technology, University of Ljubljana, and Delft University of Technology applied in 2007 to compare a large group of medium-sized European cities and evaluate their performance as smart cities. The following six domains were considered: living; economy; people; environment; mobility; and governance (Giffinger et al., 2007). This multi-dimensional intervention logic is also supported by IBM Corporation (2017a; 2017b) and Cisco Systems (2016a; 2016b) and their operating systems for smart cities: comprehensive ICT platforms that integrate a collection of digital solutions and applications for improving the management of systems for energy and utilities, parking, environmental protection, safety and security, transportation, education and healthcare. In contrast to this, the European Commission

Table 2. Capturing the dichotomous nature of smart city research in the literature published between 1992 and 2018

Hypotheses: H1.1. Techno-led; H1.2. Holistic; H2.1. Top-down; H2.2. Bottom-up; H3.1. Double Helix; H3.2. Quadruple Helix; H4.1. Mono-dimensional; H4.2. Integrated.

Reference	Dichotomy 1		Dichotomy 2		Dichotomy 3		Dichotomy 4	
	H1.1	H1.2	H2.1	H2.2	H3.1	H3.2	H4.1	H4.2
ABB, 2013	X		X		X			
Alawadhi et al., 2012				X				
Amato et al., 2012a	X		X		X			
Amato et al., 2012b	X		X		X			
Amato et al., 2012c	X		X		X			
Angelidou, 2017		X	X	X		X		
Angelidou and Psaltoglou, 2017		X						
Baccarne et al., 2014a						X		
Baccarne et al., 2014b						X		
Bergvall-Kåreborn et al., 2009			X		X			
Bolici and Mora, 2015			X	X		X		
Brech et al., 2011	X		X		X			
Breuer et al., 2014			X	X				
Caragliu et al., 2011		X						
Carvalho, 2015		X						
Chen-Ritzo et al., 2009	X		X		X			
Christopoulou et al., 2014		X						
Cisco Systems, 2016a								X
Cisco Systems, 2016b								X
Concilio and Rizzo, 2016		X						
Cosgrove et al., 2011	X		X		X			
Cugurullo, 2013		X						
Dameri, 2014						X		
Dameri, 2017						X		
Deakin, 2014		X				X		
Deakin and Al Wear, 2011				X				
Deakin and Leydesdorff, 2014						X		
Dirks and Keeling, 2009	X		X		X			
Dirks et al., 2009	X		X		X			
Dirks et al., 2010	X		X		X			
Ersoy, 2017		X						
European Commission, 2009							X	
European Commission, 2011							X	
European Commission, 2012a							X	
European Commission, 2012b							X	
European Commission, 2016							X	
European Innovation Partnership on Smart Cities and Communities, 2013							X	
Exner, 2015			X	X				
Gardner and Hespanhol, 2018							X	
Giffinger et al., 2007								X
Gooch et al., 2015				X				
Grossi and Pianezzi, 2017		X						
Harrison et al., 2010	X		X		X			
Harrison et al., 2011	X		X		X			
Hemment and Townsend, 2013		X		X				
Hollands, 2008		X						
Hollands, 2015		X						
Hollands, 2016		X						
IBM Corporation, 2017a	X		X		X			X
IBM Corporation, 2017b	X		X		X			X
Katz and Ruano, 2011	X		X		X			
Kehoe, 2011	X		X		X			
Kitchin, 2014		X						
Kohno et al., 2011	X		X		X			

(Continued)

Table 2. Continued

Reference	Dichotomy 1		Dichotomy 2		Dichotomy 3		Dichotomy 4	
	H1.1	H1.2	H2.1	H2.2	H3.1	H3.2	H4.1	H4.2
Komninos, 2014		X		X		X		
Kourtit et al., 2014		X				X		X
Kurebayashi et al., 2011	X		X		X			
Lee and Hancock, 2012			X					
Lee et al., 2014			X	X				
Leydesdorff and Deakin, 2011				X		X		
Manville et al., 2014		X						X
McNeill, 2016		X						
Mora and Bolici, 2016		X	X	X		X		
Mora and Bolici, 2017		X	X	X		X		
Niaros, 2016		X						
Paul et al., 2011	X		X		X			
Pollio, 2016		X						
Ratti and Townsend, 2011				X				
Reddy Kummitha and Crutzen, 2017		X						
Ruano et al., 2011	X		X		X			
Schaefer et al., 2011	X		X		X			
Schaffers et al., 2012				X				
Schuurman et al., 2012				X		X		
Schuurman et al., 2016				X		X		
Selada, 2017		X				X		
Shin, 2007		X		X				
Shin, 2009		X		X				
Shin and Kim, 2010		X		X				
Siemens, 2014	X		X		X			
Soderstrom et al., 2014		X				X		
Sujata et al., 2016		X						
Tamai, 2014	X		X		X			
Townsend, 2013		X		X				
van Waart et al., 2016						X		
van Winden and van den Buuse, 2017						X		
Viitanen and Kingston, 2014		X						
Yigitcanlar and Kamruzzaman, 2018		X						
Yigitcanlar and Lee, 2014		X						
Yoshikawa et al., 2011	X		X		X			
Zygiaris, 2013		X						

promotes a mono-dimensional vision of the smart city, which is described as a low-carbon and resource efficient urban environment fully committed to invest in IT solutions for smart transport, smart buildings, and smart grids.

This extensive literature review exposes the hidden contradictions of the debate on smart cities and the four dichotomies that the research conducted into the subject has generated. Each dichotomy surfaces from divergent hypotheses concerning what strategic principles drive smart city development. These hypotheses are listed in Table 3 and the scientific knowledge required to empirically test their validity can be acquired by conducting multiple-case study analyses. The following section provides a comprehensive and thorough description of the methodology that this paper proposes to organize and carry out such analyses.

A Research Methodology for Investigating Smart Cities

Case-study research involves the empirical investigation of a current phenomenon within its real-life context and can be applied to meet four different purposes: (1) to provide

Table 3. The four dichotomies emerging from smart city research and the divergent strategic principles they underpin

Dichotomies	Strategic principles
Dichotomy 1: Technology-led or holistic	**Hypothesis 1.1:** Technology-led strategy
	Hypothesis 1.2: Holistic strategy
Dichotomy 2: Top-down or bottom-up approach	**Hypothesis 2.1:** Top-down approach
	Hypothesis 2.2: Bottom-up approach
Dichotomy 3: Double or quadruple-helix model of collaboration	**Hypothesis 3.1:** Double-helix model of collaboration
	Hypothesis 3.2: Quadruple-helix model of collaboration
Dichotomy 4: Mono-dimensional or integrated intervention logic	**Hypothesis 4.1:** Mono-dimensional intervention logic
	Hypothesis 4.2: Integrated intervention logic

descriptions; (2) to build new theories; (3) to refine existing theories; and (4) to test the validity of existing theories (Eisenhardt, 1989; George and Bennett, 2005; Robson, 1993; Yin, 2009). The research methodology that this paper proposes focuses attention on the last approach: its aim is to activate a theory-testing process able to assess the validity of the divergent hypotheses emerging from each dichotomy by means of multiple-case study analyses of smart city development strategies. This methodology is built on the most relevant literature describing how case study research should be approached (Creswell, 2009; Eisenhardt 1989; Eisenhardt and Graebner, 2007; George and Bennett, 2005; Gerring, 2004; Gibbert et al., 2008; Miles and Huberman, 1994; Patton, 1990; Robson, 1993; Seawright and Gerring, 2008; Shakir, 2002; Stake, 1978, 1995, 1998; Yin, 2009; 2012) and is composed of three phases: (1) multiple-case study selection, (2) data collection, and (3) data processing and analysis.

Phase 1: Multiple-Case Study Selection

The selection of the appropriate cases is key to a successful multiple-case study analysis. Considering the purpose of this study, the selection process needed to rely on a theoretical sampling approach and not a random selection. Theoretical sampling means that case studies, as experiments conducted in a laboratory, are not randomly sampled from a population, but "chosen for the likelihood that they will offer theoretical insight" (Eisenhardt and Graebner, 2007: 27).

Starting from an initial population of cities in which a strategy for supporting smart city development has been designed and implemented, the researcher is required to find and focus attention on extreme cases: unusual manifestations of the smart city phenomenon, which show either outstanding success or failure in approaching such development. These cases are the most suitable to confirm or disprove the initial hypotheses by observing what strategic principles have driven the selected cities. The selection of extreme cases depends on how the researcher wants to approach the replication logic. The available options are the following: (1) literal replication: the cases are chosen due to their similar settings and are expected to provide similar results and (2) theoretical replication: the selected cases have different settings and variations are expected in the results of the analysis.

The first approach is based on the selection and comparison of either (1) cities that have successfully approached smart city development and the ICT-driven approach to urban sustainability it promotes, or (2) cities in which the transition process has proven to be unsuccessful. The literal replication is exemplified by Ornetzeder and Rohracher (2013), who investigate the internal dynamics and structural conditions necessary to develop successful grassroots innovations in the fields of energy and transport. The investigation is conducted by comparing three outstanding cases of sustainable grassroots innovations. Aschemann-Witzel et al. (2017) propose something similar in their investigation into key success factors in initiatives designed to reduce consumer-related food waste. Additional examples are provided by Mora and Bolici (2016; 2017) and Bolici and Mora (2015), wherein the selection of two leading cases of smart cities makes it possible to outline a preliminary roadmap describing the design and implementation process of smart city development strategies.

By following a theoretical replication logic, attention is instead focused on both types of extreme cases and the researcher tests its initial hypotheses by comparing successful and unsuccessful samples. For example, Li et al. (2012) investigate the role of public–private partnerships in residential brownfield redevelopment by analyzing two case studies that have produced opposite results. This approach has also helped Brunia et al. (2016: 30), who compare workspaces with opposite employee satisfaction scores in order to explore what factors explain "the high or low percentages of satisfied employees in offices with shared activity-based workplaces."

Both approaches are considered as suitable for this research methodology, as they are complementary in nature. However, despite the approach chosen for conducting the case study analysis, it is important to note that the multiple-case study selection process and the number of replications always determine the external validity of the analysis and the extent to which the findings can be generalized. The number of replications depends upon the certainty the researcher wants to achieve and "the greater certainty lies with the larger number of cases" (Yin, 2009: 58). Research by Eisenhardt (1989) suggests a number of cases between 4 and 10 is ideal to have a good basis for analytical generalization.[5]

In addition, analytical generalization is also affected by two contextual conditions: the geographical distribution of the selected cases and their size.[6] A more heterogeneous sample determines a broader generalization of the results. For example, Calzada (2017) investigates the governance strategies of leading smart city transformations by comparing four European cities located in two different countries: Glasgow and Bristol in the United Kingdom and Barcelona and Bilbao in Spain. This approach makes it possible to improve the common understanding of such effects in Europe, but additional research is required to test whether the findings are of wider significance because they also apply to other territorial contexts.

Phase 2: Data Collection

The researcher follows a replication logic and subjects all the selected cases to the same analytical process, which starts with the data collection phase. To establish what strategic principles have led the cities towards becoming successful or unsuccessful examples of smart cities, two databases are required. The first one includes a list of all the activities undertaken by each city to implement the smart city development strategy, to be organized and classified in four categories:

A. Community Building. Activities supporting the construction of an open and inclusive collaborative environment able to support the design and implementation of the smart city development strategy. This is done by raising citizen engagement in the smart city field; stimulating user-driven innovation and community-led urban development; increasing public awareness and digital literacy; informing the city's stakeholders; and improving their level of understanding about smart city development and the benefits it can generate.

B. Strategic Framework. Activities aiming to develop the city's strategic framework for guiding and regulating smart city development. The output of these activities includes: (1) action plans, programs, guidelines, roadmaps, recommendations, governmental acts, and policy documents; (2) measures proposing standards and technical requirements, along with assessment methods; and (3) workgroups managing the general course of the smart city development strategy's operations.

C. Services and Applications. Activities which allow new ICT services and applications to be integrated within the city.

D. Digital Infrastructure. Activities aiming to develop the technological infrastructure necessary to use and benefit from the available ICT services and applications. Examples of activities include the integration of urban operating systems and the construction or extension of high-speed broadband networks and public Wi-Fi networks.

This classification system makes it possible to group together all the activities according to the objectives and outcomes towards which efforts are directed. In case of activities producing outcomes that belong to multiple categories, they need to be included in more than one group. By analyzing the percentage of activities belonging to each group, the researcher will be able to determine whether the smart city development strategies under investigation are: (1) holistic or technology-led [Dichotomy 1] and (2) developed by means of either a top-down or bottom-up approach [Dichotomy 2].

The activities belonging to the category "[C] Services and applications" shall then be assigned to one or more application domains to investigate the smart city development strategy's intervention logic [Dichotomy 4]. These activities allow the integration of new digital solutions within the urban environment and can be classified according to the objectives pursued through their implementation. The classification system is composed of 11 application domains, which are described in Table 4 and selected by merging the classification systems for smart technologies proposed to date (Giffinger et al., 2007; Manville et al., 2014; Neirotti et al., 2014; Reviglio et al., 2013; Cisco Systems, 2016a; 2016b).[7] This typology makes it possible to build a classification system as broad as possible.

After classifying all the activities, the researcher focuses attention on the structure of each smart city development strategy's inter-organizational collaborative network [Dichotomy 3]. The aim is to build a second database describing such a network and establish whether the model of collaboration is based on a double or quadruple-helix approach. To meet this aim, the organizations that have collaborated in developing the activities previously mapped need to be identified and classified. In addition, each activity needs to be

Table 4. Classification system for application domains

Application domains	Objectives
C.01. Energy networks	To increase the efficiency and sustainability of either street lighting, or networks for producing, storing and distributing energy
C.02. Air	To ensure a better air quality in outdoor environments
C.03. Water	To improve water resource management
C.04. Waste	To improve waste management processes
C.05. Mobility and transport	To provide city users with more sustainable and accessible transport systems and address mobility issues
C.06. Buildings and districts	To improve the efficiency, accessibility and management systems of buildings and districts
C.07. Health and Social Inclusion	To improve the quality, accessibility and organization of health services and support social inclusion
C.08. Cultural heritage	To ensure a better protection of both tangible and intangible cultural heritage and enhance their cultural value
C.09. Education	To increase the quality of teaching-learning processes delivered by education systems
C.10. Public safety and security	To ensure safety and security in urban spaces and face safety challenges
C.11. E-government	To increase the convenience and accessibility of public services and information to city users
C.12. Other	ICT services and applications aiming at producing benefits different than those related to the previous application domains

analyzed to establish whether citizens have been involved in the implementation process. The following classification system is provided to group the organizations by type:

- Research: universities and other research and educational institutions
- Industry: businesses which are involved in consultancy activities and/or in the distribution of goods and services
- Government: local, regional, and national governmental authorities, along with their majority-owned subsidiaries and external agencies
- Civil Society: civil society organizations
- Other: organizations that do not belong to the previous categories or where the information necessary to complete the classification is not available.

These classification systems allow for the activities and organizations to be mapped and analyzed by cross-referencing the qualitative data extracted from multiple sources. Digital records reporting on the smart city development strategy under investigation and produced by the city government should be considered as primary sources. These include the following examples: agendas, minutes of meetings, press releases, news and newsletters, conference presentations, conference speeches obtained from either presenters' notes or videos of the events; reports, brochures, governmental acts, policy papers and documents, and webpages. Additional data can also be acquired from digital records produced by organizations that are either collaborating with the city government in implementing the city's smart city development strategy, or not involved but interested in providing data describing such a strategy. These sources can be considered secondary and can include, for example, reports produced by consultancy firms, news and articles published in online magazines, and any type of scholarly publications. This approach strengths the quality of the research process because the multiple-case study analysis is conducted by combining data and information which are extracted from multiple sources and provided by both internal and external observers.[8]

Primary and secondary sources can be found by conducting multiple keyword search queries which aim to scan the World Wide Web. The following search string is suggested: "[name of the city under investigation] smart city." If the case study is located in a non-English-speaking country, the search string should be adapted in accordance with the local language. For example, in the case of Barcelona, data items reporting on the city's smart city development strategy frequently use the term "ciudad inteligente" instead of "smart city" (City of Barcelona, 2012; 2013).

Phase 3: Data Processing and Analysis

Coding is suggested as a method to organize the large volume of unstructured qualitative data collected from the data items and facilitate the identification and classification of both activities and organizations, along with their progressive analysis.

> Coding is how you define what the data you are analyzing is about. It involves identifying and recording one or more passages of text or other data items such as the part of pictures that, in some sense, exemplify the same theoretical or descriptive idea. Usually, several passages are identified and they are then linked with a name for that idea—the code. Thus, all the text and so on that is about the same thing or exemplifies the same thing is coded with the same name. (Gibbs, 2007: 38)

The coding process can be conducted by following the procedure suggested by Eisenhardt (1989), Gibbs (2007), Robson (1993), and Strauss and Corbin (1990). Qualitative data analysis and research software programs, such as Atlas.ti, NVivo, and QDA Miner, can be deployed as supporting tools. After being collected, the digital records need to be reviewed repeatedly to identify the activities that every city that has been selected as a case study has implemented to enable smart city development. Each activity is assigned a code, identifying sections of text or other data items that describe the following attributes: objectives of the activity; generated or expected outcomes; and organizations involved in its development.[9]

The coding process is expected to result in a detailed report in which the activities of each city are listed and the data necessary to study them is presented in a structured and well-organized form. These data are then used to populate the two databases and acquire the knowledge necessary to test the divergent hypotheses emerging from the four dichotomies.

Pilot Study: Testing the Research Methodology

A small-scale preliminary study is conducted to assess the practical feasibility, effectiveness, and logistics of the proposed methodology. This makes it possible to reveal practical issues and limitations and propose changes that are able to solve them. In addition, this pilot study offers the possibility of showing researchers how the research methodology can be deployed by way of a practical example.

The pilot study is split into three phases. During the first phase, the strategy for multiple-case study selection is verified by simulating the sampling process of a multiple-case study analysis that: (1) focuses attention on large European cities (population between 500,000 and 5,000,000 inhabitants);[10] and (2) is based on a literal replication logic. The sampling process results in the identification of 10 extreme cases: large European cities that have successfully approached the implementation of strategies for supporting smart city development. One of these extreme cases is then selected to run the second and third testing phases, in which the approach proposed for collecting, processing, and analyzing the data is examined.

Testing Phase 1: Multiple-Case Study Selection

An initial population of cities in the selected range of inhabitants and belonging to the European Union's member states is defined by combining the census statistics of each country.[11] In accessing such data, 60 candidate case studies were identified, and the following were selected as extreme cases: Amsterdam in the Netherlands; London, Birmingham, Glasgow and Manchester in the United Kingdom; Copenhagen in Denmark; Barcelona and Madrid in Spain; Vienna in Austria; and Helsinki in Finland. These 10 cities were selected due to their success in the field of smart cities and heterogeneous geographical distribution. Together, they cover six different European countries, and this provides the basis for a broad generalization of the results.

The success of each candidate case study in the field of smart city development was evaluated by means of an online search phase, which was conducted to identify and review the literature reporting on: (1) comprehensive comparative analyses of smart city cases and smart city rankings in which one or more candidate case study was shown to be a leading example; and (2) competitions in which these cities have received awards for their smart city development strategies. The data resulting from this search support the identification of the ten above-mentioned extreme cases:

- After comparing infrastructural, social, and economic factors characterizing a large sample of cities, Kotkin (2009) has included Amsterdam in the top 10 world's smart cities
- Amsterdam, Copenhagen, and Vienna are respectively among the winners of the World Smart Cities Awards 2012, 2014, and 2016. In addition, along with Barcelona, Amsterdam is also one of the finalists of the 2015 edition[12]
- The Intelligent Community Forum (ICF)'s team of analysts has named Barcelona as one of the best Smart Communities of 2012, recognizing the city's leadership role in supporting urban development and innovation by leveraging the potential of ICT solutions and infrastructures[13]
- Amsterdam's smart city development strategy has been selected as a "Benchmark of Excellence" by the European Commission and described as a best practice to be replicated in other urban contexts (Velthausz, 2011)
- Amsterdam was awarded the European City Star Award 2011 by the European Commission, which highlighted the capability of its smart city development strategy to demonstrate how cities can be successful in bringing together public parties, private organizations, and citizens in order to take advantage of ICT for urban development purposes (I amsterdam, 2011; Amsterdam Smart City, 2011)
- According to the Smart City Index Rankings developed in 2011 and 2012 by IDC (International Data Corporation), Barcelona and Madrid are both among the top five smart cities in Spain (Achaerandio et al., 2011; 2012)
- In a recent study benchmarking Austrian smart cities, IDC (2016) has also included Vienna among the most advanced smart city cases

- The European Commission nominated Barcelona as the 2014 European Capital of Innovation for its smart city development strategy. According to the jury, this strategy showed how the use of ICT could bring the city government closer to its citizens (European Commission, 2014a)
- According to a new study commissioned by Huawei and conducted by Navigant Consulting (Woods et al., 2016), London, Birmingham, Glasgow, and Manchester are the United Kingdom's leading examples of smart cities
- Vienna heads the Smart City Strategy Index developed by the consulting firm Roland Berger, while Stockholm and Copenhagen are among the top-performing European cities (Zelt et al., 2017)
- Considering the data provided by Manville et al. (2014), the Directorate-General for Internal Policies of the European Parliament has recognized Amsterdam, Barcelona, Copenhagen, Helsinki, Manchester, and Vienna as six of the most successful smart cities in Europe and the most suitable cases for further in-depth analyses.

Testing Phase 2: Data Collection

The practical feasibility and effectiveness of the data collection, processing, and analysis processes were tested by using the case of Vienna, which was randomly selected among the 10 extreme cases. This made it possible to continue with the pilot study and start searching for the digital records from which to collect the qualitative data necessary to conduct the analysis.

The data collection process was composed of a series of searches,[14] each one pursuing a specific aim, in which Google and multiple search strings were deployed:

- Search 1: "Vienna smart city" OR "Wien smart city"
- Search 2: "Vienna smart city" OR "Wien smart city" site:smartcity.wien.gv.at
- Search 3: "Vienna smart city" OR "Wien smart city" site:www.wien.gv.at
- Search 4: "Vienna smart city" OR "Wien smart city" -"www.wien.gv.at" -"smartcity.wien.gv.at"

Search 1 was conducted in order to find the main online repositories in which the city government of Vienna stores the digital records reporting on the city's smart city development strategy. The pages displayed by the search engine in response to the query show that the city government's repositories storing most of the data items are the City Council's online information service (https://www.wien.gv.at) and *Smart City Wien*, the official website of the Vienna's smart city development strategy (https://smartcity.wien.gv.at).

Both repositories were searched (Search 2 and Search 3) and this made it possible to detect 365 digital records, which included press releases, news, newsletters, webpages, interviews, conference presentations, reports, posts on social media websites, policy papers and documents, and governmental acts. After being identified, every data item was downloaded and labelled using consecutive numbers. In addition, an excel spreadsheet is created in which the items' Uniform Resource Locators (URLs) were listed in order to check the presence of duplicates, which were eliminated as soon as detected.

This list was then expanded upon by adding 114 new digital records produced by organizations that were either collaborating with the city government in implementing Vienna's smart city development strategy or interested in communicating information describing the program of activities that were undertaken. These organizations were consultancy firms, publishing companies, research centers, universities, national and regional governmental authorities, and non-governmental institutions, and the digital records they produced include scholarly publications, articles found in online magazines and newspapers, posts, press releases, reports, and videos. These data items were considered as secondary sources and were collected with Search 4, in which Google was asked to automatically eliminate any results from the two city governments' online repositories previously examined.

Testing Phase 3: Data Processing and Analysis

Overall, 99.3 percent of the collected digital records were text documents, while the remaining items were digital videos. Considering the high number of data items, their analysis was conducted by using *Atlas.ti* as a supporting tool. *Atlas.ti* works with a large range of media; however, in order to be processed, files need to be formatted according to the system's requirements. Therefore, text documents stored in formats different from .txt, .doc, .docx, .odt, and .pdf were all converted. However, no changes were required for the video files.

After being prepared and uploaded onto *Atlas.ti*, the digital records were reviewed in a systematic way to identify the activities that Vienna has developed to implement its smart city development strategy. Each activity was assigned a code that described the following attributes of the data: objectives of the activity; generated or expected outcomes; and entities involved in its development. The coding process resulted in a detailed report in which the activities were listed one by one and the data necessary to study them were presented in a structured and well-organized form. The report was generated by using *Atlas.ti*'s output function called Codebook, and was used to create two databases, in which activities and organizations were classified (see Appendix B and Appendix C). It is important to note that the information necessary to classify each organization was obtained from their official websites, because the data provided by the digital records were insufficient to complete this task.

With the coding process, 54 activities were mapped (See Figure 1) and their analysis makes it possible to understand how Vienna has approached smart city development and test the validity of the hypotheses each dichotomy stands on.

Dichotomy 1: Technology-Led or Holistic Strategy

Vienna's smart city development strategy gives equal weight to: (1) the deployment of technological advancements leading to either the resolution or mitigation of urban sustainability issues; and (2) the development of both a collaborative environment and a strategic framework for supporting the deployment of these technological advancements. This strategy is therefore based on a holistic vision of smart cities, which are not considered as technology-only focused systems resulting from the massive combination of sets of interconnected ICT components, but socio-technical systems in which technological development is aligned with human, social, cultural, economic, and environmental factors.

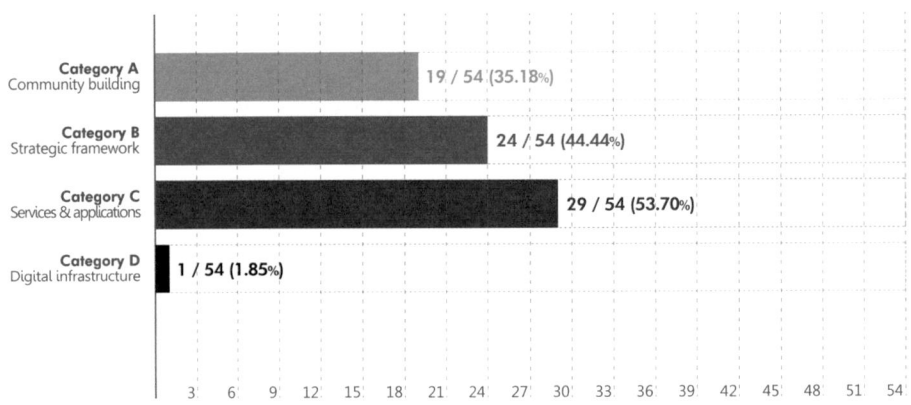

Figure 1. Vienna's smart city development strategy: number of activities by category

The accuracy of this statement is evidenced by the data in Figure 2, in which the percentage of activities by group of categories is compared and appears to be balanced. The first group includes those activities belonging to at least one of the first two categories, i.e., "[A] Community building" and "[B] Strategic framework," both of which focus attention on the non-technological factors of smart city development. For example, with the European projects CLUE, TRANSFORM, and Urban Learning, Vienna has improved its capability of delivering policy and programs for supporting the deployment of ICT solutions able to reduce carbon emissions (Brandt et al., 2014; Hartmann et al., 2015; Hemis et al., 2017; CLUE Project Partners, 2014). In collaboration with a consortium composed of 28 partners, which includes city governments, businesses, and universities, Vienna has also launched Smart Together, a project aimed at testing new approaches for fostering user-centric innovation, collaboration, and citizen engagement in smart city development.[15]

On the contrary, the second group of activities include projects and initiatives in which the deployment of ICT services, applications, and infrastructures within the urban environment is among the objectives or the only objective. Examples of technological solutions and infrastructures include: decision-supporting tools for managing urban energy and mobility systems (Bednar et al., 2016; Marguerite et al., 2016); electric vehicles, charging infrastructures and info-mobility systems (Wiener Modellregion and Climate and

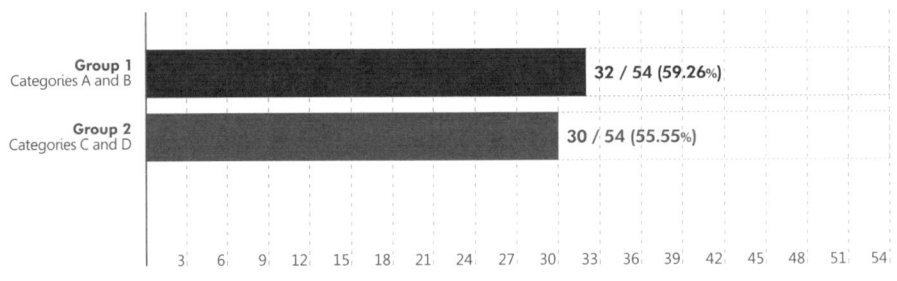

Categories: **A** Community building; **B** Strategic framework; **C** Services and applications; **D** Digital infrastructure

Figure 2. Vienna's smart city development strategy: number of activities by group of categories

Energy Fund, 2014); a large-scale network of Wi-Fi access points in public spaces and leisure areas to provide citizens and tourists with location-based information and free-of-charge access to the Internet; mobile apps allowing the city government to receive feedback and send up-to-date information and instructions on how to proceed in case of dangerous situations; and QR codes for accessing digital contents related to local facilities via mobile devices.[16]

Dichotomy 2: Top-Down or Bottom-Up Approach

Vienna's smart city development strategy is holistic and keeps a balance between top-down and bottom-up approaches. The city government is the most active organization belonging to the smart city ecosystem of Vienna and has contributed to develop about 50 percent of the total activities (See Appendix C). This means that Vienna's smart city development is boosted by a significant number of bottom-up activities and, what is more, the analysis of the objectives and outcomes related to the work undertaken by the city government demonstrates that it is clearly aimed at promoting this bottom-up development process. The city government provides leadership and its actions are oriented towards the construction of: (1) a decentralized development process; (2) an open, inclusive, and cohesive collaborative ecosystem; and (3) the strategic framework for regulating the smart city transformation of the entire city and bringing the different organizations belonging to this ecosystem into a harmonious and efficient relationship.

To achieve this aim, the city government:

- sets up a participatory process for developing the Smart City Wien Framework Strategy, i.e., a strategic document that lays down Vienna's guidelines for smart city development. This document provides a long-term vision that extends to 2050 and establishes what objectives need to be achieved and the expected results. In addition, it identifies the application domains to focus attention on and describes the governance and monitoring systems that need to be adopted and the strategic principles to follow. The use of a participatory approach ensures the Framework Strategy represents a single vision that city stakeholders all agree on (City of Vienna 2014)
- collects ideas, comments, and feedback about the city's ICT requirements from public and private sector organizations and citizens in order to develop a Digital Agenda with projects for handling Vienna's most pressing urban challenges (Heissenberger and Schuhböck, 2015)
- increases Vienna's know-how on urban technologies and smart city development by collaborating in delivering new planning and operational tools, recommendations, guidelines, standards and technical requirements, and evaluation and assessment methods[17]
- assigns the role of Smart City Wien Agency to TINA Vienna GmbH,[18] which becomes "the central coordination point for all internal and external stakeholders. It should cover the areas of coordination, stakeholder management, inquiry management, and communication and would record, evaluate, and initiate projects on behalf of all relevant partners within and outside the City of Vienna. The objective lies in the interdisciplinary promotion of networking between municipal administration, research, business, and industry" (City of Vienna, 2014: 88)

- collaborates in organizing forums, conferences, workshops, and meetings dealing with smart city development in order to: generate interest; inform the community; engage new stakeholders and make the collaborative ecosystem larger; stimulate collaboration; and raise public awareness of the potential benefits ICTs can produce in urban environments (City of Vienna, 2016; Digital City Wien 2015; 2016)
- makes public data freely accessible to support developers interested in building new applications and digital services.[19]

Dichotomy 3: Double or Quadruple-Helix Model of Collaboration

The smart city collaborative ecosystem of Vienna is analyzed and graphically visualized by using the open-source software Gephi. The result is the network illustrated in Figure 3, in which the organizations mapped during the coding process are represented as nodes with a diameter that is directly proportional to the number of activities they have worked on. Every edge connects the organizations which have collaborated in implementing at least one activity. The stronger the degree of collaboration between two organizations, the higher the thickness of the edge connecting them. Colors are assigned according to the organization types.

The data describing the networks' structure shows that Vienna has approached smart city development by means of a triple-helix collaborative model: the collaboration among industry, government, and research is the engine behind Vienna's smart city development strategy. With participation at 56 percent, businesses are the most active organizations and are followed by institutions for education and research and governmental authorities, which both represent approximately 19 percent of the collaborative network. The remaining 4 percent are civil society organizations, which are by far the least represented organization type.

However, despite this data, it is important to note that a number of activities suggest Vienna has made an effort to strengthen the participation possibilities of civil society by increasing citizens' active involvement in the implementation process of its smart city development strategy (See Figure 4). This intent is clearly expressed in the strategic frameworks that the city government has developed to guide and regulate the development of Vienna as a smart city (City of Vienna, 2014). For example, the city government has invited Vienna's citizens to take part in the planning phase of the smart city development strategy and support the production of the strategic framework by attending a series of stakeholders' forums, along with representatives from universities, governmental authorities, the business sector, and civil society organizations. These forums have been organized regularly and conceived as discussion events for exchanging ideas and ensuring greater transparency, participation, and collaboration in the smart city field (Hofstetter and Vogl, 2011; Climate and Energy Fund, 2013; City of Vienna et al., 2011; 2013; City of Vienna, 2012; 2013a; 2013b; 2014; 2016).

In addition, a number of projects have activated collaborative processes in which citizens are asked to participate in the design and testing phases of ICT solutions and infrastructures to be deployed in the urban environment. The aim is to improve the capability of ICT to meet local communities' needs. This is the case of Smarter Together and Smart Cities Demo Aspern, in which Living Labs are used as collaborative platforms for attracting high public attention and foster user-centric innovation (Aspern Smart City Research,

2015; Muhlmann, 2017).[20] All of this provides evidence of an attempt to move from a triple to a quadruple-helix collaborative model.

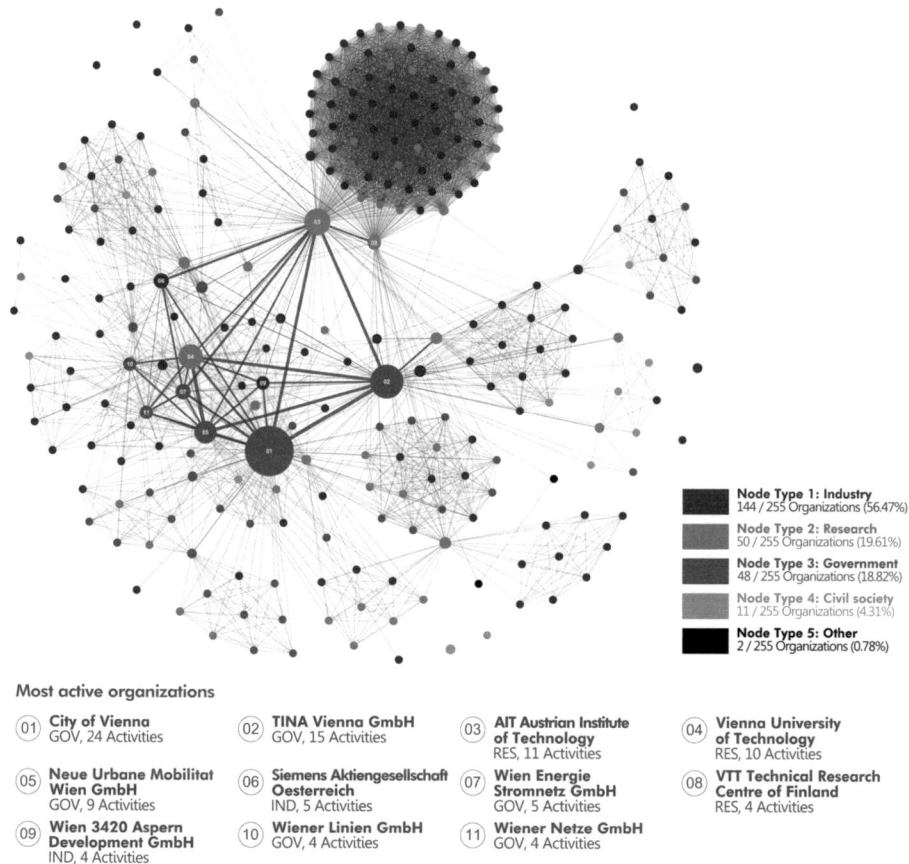

Most active organizations

01 **City of Vienna** GOV, 24 Activities	02 **TINA Vienna GmbH** GOV, 15 Activities	03 **AIT Austrian Institute of Technology** RES, 11 Activities	04 **Vienna University of Technology** RES, 10 Activities
05 **Neue Urbane Mobilitat Wien GmbH** GOV, 9 Activities	06 **Siemens Aktiengesellschaft Oesterreich** IND, 5 Activities	07 **Wien Energie Stromnetz GmbH** GOV, 5 Activities	08 **VTT Technical Research Centre of Finland** RES, 4 Activities
09 **Wien 3420 Aspern Development GmbH** IND, 4 Activities	10 **Wiener Linien GmbH** GOV, 4 Activities	11 **Wiener Netze GmbH** GOV, 4 Activities	

Figure 3. Vienna's smart city development strategy: smart city collaborative ecosystem

Dichotomy 4: Mono-Dimensional or Integrated Intervention Logic

The data obtained from the analysis of the activities belonging to the category "[C] Services and applications" shows that Vienna has adopted an integrated intervention logic

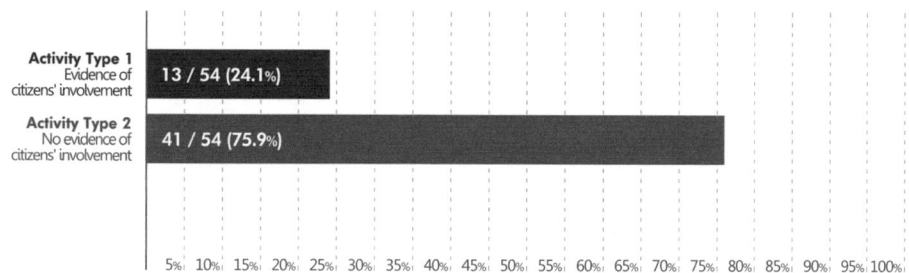

Figure 4. Vienna's smart city development strategy: citizen participation

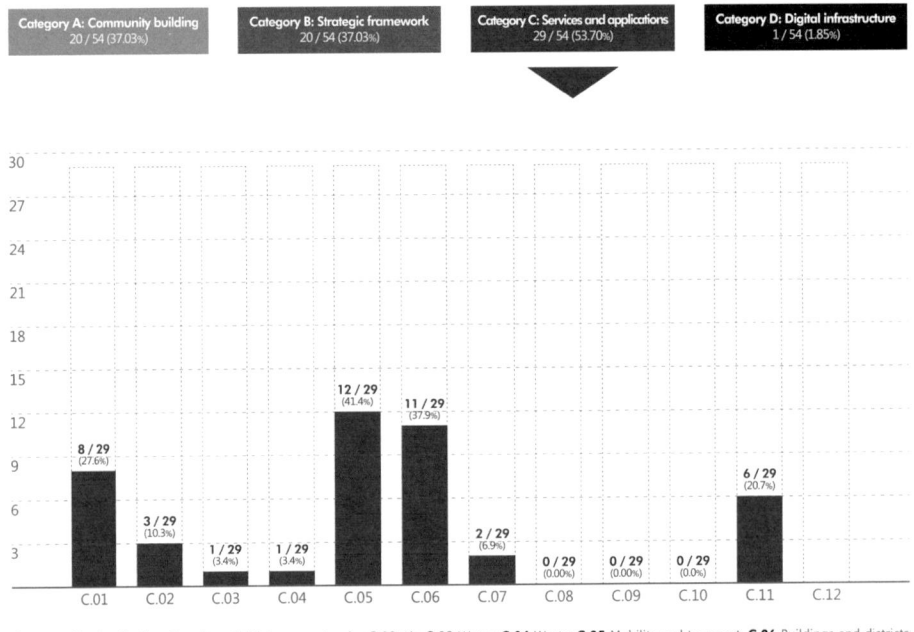

Figure 5. Vienna's smart city development strategy: activities by application domain

and its smart city development strategy covers a mix of application domains (See Figure 5). The city's interest in smart city development is mainly oriented towards smart transport, smart building, and smart grid solutions for low-carbon and energy-efficient urban environments. Most of the ICT services and applications supporting Vienna's smart city transformation are deployed to fight climate change and boost energy efficiency in mobility and transport (C.05), buildings and city districts (C.06), and power infrastructures (C.01).

This approach is aligned with the European Commission's interpretation of smart cities. However, Vienna's smart city development strategy makes a significant effort to extend such an interpretation by seeking to use digital technologies for addressing additional sustainability issues related to other policy domains. For example, technological solutions are brought into action to improve the management of natural resources other than energy, such as air, waste, and water; stimulate the use of public transports by providing citizens with real-time information; stimulate social inclusion and citizen collaboration by offering up-to-date overviews of where co-design activities are implemented in Vienna and opening public data; and increase the quality of assistance services for elderly people by helping them to easily access the online world.[21]

Discussion and Conclusions

Overcoming the dichotomous nature of smart city research is fundamental in order to provide cities with a clear understanding of how smart city development should be approached, and to support them in delivering the program of activities needed to enable such a development. This paper introduces a research methodology for conducting the deductive multiple-case study analyses necessary to meet this challenge. In addition, it reports on the practical feasibility, effectiveness, logistics, and replicability of this methodology by analyzing the implementation process of the smart city development strategy proposed by Vienna.

The results of the pilot study show how the proposed research methodology can be deployed to capture and codify the knowledge produced from multiple smart city experiences. This common protocol makes it possible for smart city researchers to (1) coordinate efforts in investigating which strategic principles drive smart city development and test the divergent hypotheses emerging from the scientific literature; (2) share the results of this investigation and hypothesis-testing by conducting extensive cross-case analyses among multiple studies able to capture the generic qualities of the findings; and (3) collaborate in developing a platform able to generate agreement over the way to think about, conceptualize, and standardize the analysis of smart city developments. The standard of reporting that this methodology lays down also makes it possible to account for and make transparent the transformation that cities undergo in the process of becoming smart. This makes it easier for researchers to capitalize on the lessons learned from extreme cases and develop new and innovative monitoring and evaluation systems for assessing the operational value of smart city development strategies.

This study clearly demonstrates that smart cities emerge from a program of associated activities and, by adopting a reverse engineering approach, these programs can be deconstructed into building blocks whose analysis enables the structure of any smart city development strategy to be interpreted. This is the rationale behind the research methodology for multiple-case study analyses that this paper reports on and whose validity, effectiveness, and replicability are demonstrated by the results of the pilot study.

The pilot study reveals the strategic principles that Vienna has decided to choose in order to enable smart city development. These principles have allowed Vienna to first address the dichotomous nature of smart city research and overcome the ambiguity surrounding the methodology driving smart city development. This is achieved by:

(1) embracing a holistic vision of smart cities, which are considered as socio-technical systems in which technological development is aligned with human, social, cultural, economic, and environmental factors
(2) balancing top-down and bottom-up approaches
(3) instituting the industry-government-research relationships of the triple-helix collaborative model, while making efforts to strengthen the participation of civil society and progressively moving towards a quadruple-helix model of stakeholder engagement
(4) assembling an integrated intervention logic that cuts across the multitude of application domains which smart city development represents.

Finally, in terms of logistics, the pilot study clearly shows that the sampling approach, data collection, and analytics of the protocol are all adequate, as is the classification system of the smart city activities that has been deployed. It also demonstrates that samples are available when looking for extreme cases of smart cities, and the literature reporting on development strategies can provide a great deal of qualitative data able to support comprehensive analytical processes. In addition, it is important to point out that no relevant issues serving to raise questions as to the value of the protocol have been detected when conducting the pilot study. However, researchers need to be aware that the coding process is particularly time-consuming and resource intensive, because it requires a systematic analysis of qualitative data which is not only large in scale, but also complex to approach in terms of content analysis. The use of digital supporting tools (*Atlas.ti*) has been fundamental to managing this phase of the analysis and maintaining a chain of evidence supporting the progressive identification of new activities. The contribution of multiple researchers in coding the activities is also extremely helpful when conducting the content analysis, because this reciprocal exercise makes it possible to check the data items in rotation and improve the quality of the coding process.

Notes

1. In this article, smart city development is considered as an ICT-driven approach to urban sustainability, and smart cities are defined as follows: cities in which issues limiting sustainable urban development are tackled by means of ICT-related solutions (Mora et al., 2018b). The term "city" is here used to refer to any type of urban area, irrespective of its population size.
2. Case study research can be approached by using either single- or multiple-case studies. However, as Yin (2009) and Eisenhardt (1989) suggest, multiple-case study analyses provide broader opportunities to generalize the theoretical prepositions under investigation than examinations based on single-case studies.
3. The quadruple-helix collaborative model is driven by the university-industry-government relations composing the triple-helix model (Etzkowitz and Leydesdorff, 2000) and adds civil society, i.e., civil society organizations and citizens, as the fourth element of the collaborative ecosystem (Arnkil et al., 2010; Cavallini et al., 2016).
4. United Nations (1998), Food and Agriculture Organization (2013), World Economic Forum (2013), and European Commission (2014b) define civil society organizations as a broad category that encompasses non-governmental organizations (NGOs), community groups, charities, trusts, foundations, advocacy groups, faith-based organizations, and national and international non-state associations.
5. Case study research does not allow for statistical generalization but analytical generalization. As Yin (2009: 38) points out, in case study research "the mode of generalization is analytic generalization, in which a previously developed theory is used as a template with which to compare the empirical results of the case study. If two or more cases are shown to support the same theory, replication may be claimed. The empirical results may be considered yet more potent if two or more cases support the same theory but do not support an equally plausible, rival theory."
6. For the selection of the cases, we suggest using the classification system developed by the European Commission and Organization for Economic Co-operation and Development (OECD), in which cities are divided by number of inhabitants. The classification system is composed of the following six categories: S (50,000–100,000); M (100,000–250,000); L (250,000–500,000); XL (500,000–1,000,000); XXL (1,000,000–5,000,000); Global cities (More than 5,000,000) (Dijkstra and Poelman, 2012).
7. See also IBM: http://www-03.ibm.com/ibm/history/ibm100/us/en/icons/smarterplanet.
8. Interviews can be considered as an additional source of information, but it is important to remember that this data collection method requires a great deal of resources and preparation; key informants need to be selected with precision and a protocol needs to be designed to guarantee a high degree of data reliability (Robson, 1993; Yin, 2009). Therefore, we suggest conducting interviews only when indispensable, in particular whether: (1) digital records are scarce or do not provide sufficient data on the program of activities of the

smart city development strategies under investigation, or (2) further validation of data through cross verification is needed.

9. Data items different from texts include pictures, maps, technical drawings, and specific sections of audio and video files.

10. In the classification system developed by European Commission and OECD, cities with a population between 500,000 and 5 million inhabitants belong to the XL and XXL categories (Dijkstra and Poelman, 2012).

11. The census statistics are presented in the Appendix A.

12. The World Smart Cities Awards is a competition that the Smart City Expo World Congress organizes annually to identify the most ambitious strategies, the most advanced projects, and the most innovative initiatives around the world fostering the development of the smart city concept. The complete lists of winners and finalists of each edition of the World Smart Cities Awards are available on the Smart City Expo World Congress' website: http://www.smartcityexpo.com.

13. The Intelligent Community Forum is an independent think tank based in New York City. All the information on its Awards Program and the winners of each edition can be found at: http://www.intelligentcommunity.org.

14. The search phase was conducted in April 2016.

15. Additional information can be found in the project's official website: http://smarter-together.eu.

16. Additional data on mobile applications and Wireless LAN hotspots currently available in Vienna is provided by the city government: https://www.wien.gv.at.

17. See the following projects: CityOpt (http://www.cityopt.eu); TRANSFORM (http://urbantransform.eu); PUMAS (www.pumasproject.eu); TRANSFORM+ (http://www.transform-plus.at); EU-GUGLE (http://eu-gugle.eu); and INNOSPIRIT (http://www.innospirit.org).

18. TINA Vienna GmbH is part of Wien Holding GmbH (2012), a holding company of the City of Vienna that carries out community tasks.

19. Data are provided through the online platform Open Government Wien (https://open.wien.gv.at/site/open-data/).

20. For further information, see also (1) the website of the Austrian Institute of Technology (AIT), which has collaborated in delivering the project Smart Cities Demo Aspern (https://www.ait.ac.at/en/research-fields/smart-grids/projects/smart-cities-demo-aspern); (2) the official website of the project Smarter together (http://smarter-together.eu).

21. See the following projects: CO2 Neutrale; SternE; SeniorTab; Citybike Wien; E-Taxis; SMILE; AnachB; Wien.at live-App; Open Government Data; and Wien Gestalten (https://smartcity.wien.gv.at).

Disclosure Statement

No potential conflict of interest was reported by the authors.

Bibliography

ABB, *ABB Power and Automation: Solid Foundations for Smart Cities* (Zurich: ABB, 2013) <http://new.abb.com/docs/default-source/smart-grids-library/abb_smart_grids_white_paper_2013.pdf?sfvrsn=2> Accessed April 5, 2014.

ARUP, *Global Innovators: International Case Studies on Smart Cities* (London: Government of the United Kingdom Department for Business, Innovation and Skills, 2013) <https://www.gov.uk/government/uploads/system/uploads/attachment_data/file/249397/bis-13-1216-global-innovators-international-smart-cities.pdf> Accessed September 8, 2016.

R. Achaerandio, G. Gallotti, J. Curto, R. Bigliani, and F. Maldonado, *Smart Cities Analysis in Spain* (Framingham, MA: IDC, 2011) <http://www.idc.com/getdoc.jsp?containerId=EIRS56T> Accessed August 6, 2014.

R. Achaerandio, J. Curto, R. Bigliani, and G. Gallotti, *Smart Cities Analysis in Spain 2012: The Smart Journey* (Framingham, MA: IDC, 2012) <http://www.portalidc.com/resources/white_papers/IDC_Smart_City_Analysis_Spain_EN.pdf> Accessed August 6, 2014.

Y.A. Aina, "Achieving Smart Sustainable Cities with GeoICT Support: The Saudi Evolving Smart Cities," *Cities: The International Journal of Urban Policy and Planning* 71 (2017) 49–58.

S. Alawadhi, A. Aldama-Nalda, H. Chourabi, R.J. Gil-Garcia, S. Leung, S. Mellouli, T. Nam, T.A. Pardo, H.J. Scholl, and S. Walker, "Building Understanding of Smart City Initiatives," in H.J. Scholl, M. Janssen, M.A. Wimmer, C.E. Moe, and L.S. Flak, eds, *Electronic Government: 11th IFIP WG 8.5 International Conference, EGOV 2012, Kristiansand, Norway, September 3–6, 2012. Proceedings* (Berlin: Springer, 2012) 40–53.

V. Amato, L. Bloomer, A. Holmes, and S. Kondepudi, *Government Competitiveness Drives Smart +connected Communities Initiative* (San Jose, CA: Cisco Systems, 2012a) <http://www.smartconnectedcommunities.org/docs/DOC-2174> Accessed January 4, 2013.

V. Amato, L. Bloomer, A. Holmes, and S. Kondepudi, *Using ICT to Deliver Benefits to Cities by Enabling Smart+Connected Communities* (San Jose, CA: Cisco Systems, 2012b) <http://www.smartconnectedcommunities.org/docs/DOC-2150> Accessed January 4, 2013.

V. Amato, S. Kondepudi, and A. Holmes, *Transforming Communities with Smart+connected Services* (San Jose, CA: Cisco Systems, 2012c) <http://www.smartconnectedcommunities.org/docs/DOC–2129> Accessed January 4, 2013.

Amsterdam Smart City, *Smart Stories* (Amsterdam: Amsterdam Smart City, 2011) <https://issuu.com/amsterdamsmartcity/docs/smart_stories> Accessed August 5, 2016.

J. Anderson, D. Fisher, and L. Witters, *Getting Smart About Smart Cities: Understanding the Market Opportunity in the Cities of Tomorrow* (Paris: Alcatel–Lucent, 2012) <http://www2.alcatel-lucent.com/knowledge-center/admin/mci-files-1a2c3f/ma/Smart_Cities_Market_opportunity_Market Analysis.pdf> Accessed January 1, 2013.

M. Angelidou, "Smart City Policies: A Spatial Approach," *Cities: The International Journal of Urban Policy and Planning* 41:Supplement 1 (2014) S3–S11.

M. Angelidou, "The Role of Smart City Characteristics in the Plans of Fifteen Cities," *Journal of Urban Technology* 24:4 (2017) 3–28.

R. Arnkil, A. Järvensivu, P. Koski, and T. Piirainen, *Exploring Quadruple Helix: Outlining User-oriented Innovation Models* (Tampere: University of Tampere, 2010) <https://tampub.uta.fi/bitstream/handle/10024/65758/978-951-44-8209-0.pdf?sequence=1> Accessed July 10, 2016.

J. Aschemann-Witzel, I.E. de Hooge, H. Rohm, A. Normann, M. Bonzanini Bossle, A. Grønhøj, and M. Oostindjer, "Key Characteristics and Success Factors of Supply Chain Initiatives Tackling Consumer–related Food Waste – A Multiple Case Study," *Journal of Cleaner Production* 155 (2017) 33–45.

Aspern Smart City Research, *Aspern Smart City Research: Energieforschung Gestaltet Energiezukunft* (Vienna: Aspern Smart City Research, 2015) <http://www.ascr.at/wp-content/uploads/2015/09/ASCR_Folder_dt.pdf> Accessed May 3, 2017.

B. Baccarne, P. Mechant, and D. Schuurman, "Empowered Cities? An Analysis of the Structure and Generated Value of the Smart City Ghent," in R.P. Dameri and C. Rosenthal-Sabroux, eds, *Smart*

City: How to Create Public and Economic Value with High Technology in Urban Space (Cham: Springer, 2014a) 157–182.

B. Baccarne, D. Schuurman, P. Mechant, and L. De Marez, "The Role of Urban Living Labs in a Smart City," in *XXV ISPIM Innovation Conference: Innovation for Sustainable Economy and Society* (Manchester: International Society for Professional Innovation Management, 2014b).

T. Bakici, E. Almirall, and J. Wareham, "A Smart City Initiative: The Case of Barcelona," *Journal of the Knowledge Economy* 4:2 (2013) 135–148.

T. Bednar, D. Bothe, J. Forster, S. Fritz, N. Haufe, T. Kaufmann, P. Eder-Neuhauser, P. Pfaffenbichler, N. Rab, J. Schleicher, G. Weinwurm, C. Winkler, and M. Ziegler, *URBEN–DK: Results Report* (Vienna: TU Wien, 2016) <https://urbem.tuwien.ac.at/fileadmin/t/urbem/files/URBEM_Ergebnisbericht_Einzelseiten_EN.pdf> Accessed August 5, 2017.

B. Bergvall-Kåreborn, C. Ihlström Eriksson, A. Ståhlbröst, and J. Svensson, "A Milieu for Innovation: Defining Living Labs," in *Proceedings of the 2nd ISPIM Innovation Symposium* (Manchester: International Society for Professional Innovation Management [ISPIM], 2009).

R. Bolici and L. Mora, "Urban Regeneration in the Digital Era: How to Develop Smart City Strategies in Large European Cities," *TECHNE: Journal of Technology for Architecture and Environment* 5:2 (2015) 110–119.

N. Brandt, F. Cambell, M. Deakin, S. Johansson, M. Malmström, K. Mulder, U. Pesch, H. Shahrokni, O. Tatarchenko, and L. Årman, *European Cities Moving Towards Climate Neutrality* (2014) <http://www.clue-project.eu/getfile.ashx?cid=69201&cc=5&refid=6> Accessed July 8, 2017.

B. Brech, R. Rajan, J. Fletcher, C. Harrison, M. Hayes, J. Hogan, L. Hopkins, P.K. Isom, J. Meegan, C. Penny, J.L. Snowdon, and D.A. Wood, *Smarter Cities Series: Understanding the IBM Approach to Efficient Buildings* (Armonk, NY: IBM Corporation, 2011) <http://www.redbooks.ibm.com/redpapers/pdfs/redp4735.pdf> Accessed September 14, 2012.

J. Breuer, N. Walravens, and P. Ballon, "Beyond Defining the Smart City: Meeting Top-down and Bottom-up Approaches in the Middle," *TeMA: Journal of Land Use, Mobility and Environment* 7 (2014) 153–164.

S. Brunia, I. De Been, and T.J. van der Voordt, "Accommodating New Ways of Working: Lessons from Best Practices and Worst Cases," *Journal of Corporate Real Estate* 18:1 (2016) 30–47.

I. Calzada, "The Techno-Politics of Data and Smart Devolution in City-Regions: Comparing Glasgow, Bristol, Barcelona, and Bilbao," *Systems* 5:1 (2017).

A. Caragliu, C. Del Bo, and P. Nijkamp, "Smart Cities in Europe," *Journal of Urban Technology* 18:2 (2011) 65–82.

P. Cardullo and R. Kitchin, *Being a Citizen in the Smart City: Up and Down the Scaffold of Smart Citizen Participation* (Maynooth: Maynooth University, 2017) <https://osf.io/preprints/socarxiv/v24jn> Accessed September 12, 2017.

L. Carvalho, "Smart Cities from Scratch? A Socio-technical Perspective," *Cambridge Journal of Regions, Economy and Society* 8:1 (2015) 43–60.

S. Cavallini, R. Soldi, J. Friedl, and M. Volpe, *Using the Quadruple Helix Approach to Accelerate the Transfer of Research and Innovation Results to Regional Growth* (EU Committee of the Regions, 2016) <http://cor.europa.eu/en/documentation/studies/Documents/quadruple-helix.pdf> Accessed November 8, 2016.

C. Chen-Ritzo, C. Harrison, J. Paraszczak, and F. Parr, "Instrumenting the Planet," *IBM Journal of Research and Development* 53:3 (2009) 338–353.

E. Christopoulou, D. Ringas, and J. Garofalakis, "The Vision of the Sociable Smart City," in N. Streitz and P. Markopoulos, eds, *Distributed, Ambient, and Pervasive Interactions: Second International Conference, DAPI 2014, Held as Part of HCI International 2014, Heraklion, Crete, Greece, June 22–27, 2014. Proceedings* (Berlin: Springer, 2014) 545–554.

Cisco Systems, *Smart Cities Exposé: 10 Cities in Transition* (San Jose, CA: Cisco Systems, 2012) <http://www.pageturnpro.com/Cisco/41742-Smart-Cities-Expose-10-Cities-in-Transition/index.html#44> Accessed January 5, 2013.

Cisco Systems, *Cisco Smart+Connected Digital Platform: At-a-glance* (San Jose, CA: Cisco Systems, 2016a) <http://www.cisco.com/c/dam/en_us/solutions/industries/docs/at-a-glance-c45-736521.pdf> Accessed June 20, 2017.

Cisco Systems, *Cisco Smart+Connected Digital Platform: Data Sheet* (San Jose, CA: Cisco Systems, 2016b) <http://www.cisco.com/c/dam/en_us/solutions/industries/docs/datasheet-c78-737127.pdf> Accessed June 20, 2017.

City of Barcelona, *Compromiso Ciudadano Por La Sostenibilidad 2012–2022* (Barcelona: Ayuntament de Barcelona, 2012) <http://ajuntament.barcelona.cat/ecologiaurbana/sites/default/files/Compromiso%20Ciudadano%20por%20la%20Sostenibilidad.pdf> Accessed August 5, 2017.

City of Barcelona, *Smart Cities: Informe Sectorial 2013* (Barcelona: Ayuntament de Barcelona, 2013) <https://bcnroc.ajuntament.barcelona.cat/jspui/bitstream/11703/86798/1/13754.pdf> Accessed August 5, 2017.

City of Vienna, *Smart City Wien Stakeholder Forum: Wo Stehen Wir* (Vienna: City of Vienna, 2012) <https://www.wien.gv.at/stadtentwicklung/studien/pdf/b008327.pdf> Accessed September 1, 2016.

City of Vienna, *Smart City Wien Stakeholder Forum: Auf Dem Weg Zur Smart City Wien Rahmenstrategie* (Vienna: City of Vienna, 2013a) <https://www.wien.gv.at/stadtentwicklung/studien/pdf/b008381.pdf> Accessed September 1, 2016.

City of Vienna, *Smart City Wien Stakeholder Forum: Innovation Durch Smarte Projekte* (Vienna: City of Vienna, 2013b) <https://www.wien.gv.at/stadtentwicklung/studien/pdf/b008328.pdf> Accessed September 1, 2016.

City of Vienna, *Smart City Wien: Framework Strategy* (Vienna: City of Vienna, 2014) <https://smartcity.wien.gv.at/site/files/2014/09/SmartCityWien_FrameworkStrategy_english_doublepage.pdf> Accessed August 30, 2016.

City of Vienna, *Stakeholder-Prozesse: Smart City Wien* (Vienna: City of Vienna, 2016) <https://www.wien.gv.at/stadtentwicklung/projekte/smartcity/stakeholder.html> Accessed September 1, 2016.

City of Vienna, 3420 Aspern Development AG, Siemens AG Österreich, Österreichisches Forschungs- und Prüfzentrum Arsenal GesmbH, raum & kommunikation GmbH, Vienna University of Technology, Energieinstitut der Wirtschaft GmbH, and Austrian Institute of Technology GmbH, *Smart City Wien: Short Report* (Vienna: Climate and Energy Fund, 2011) <http://www.smartcities.at/assets/Projektberichte/Endbericht-Kurzfassung/Endbericht-K11NE2F00030-Wien-kurz-dt-engl-v1.0.pdf> Accessed September 1, 2016.

City of Vienna, 3420 Aspern Development AG, Siemens AG Österreich, Österreichisches Forschungs- und Prüfzentrum Arsenal GesmbH, raum & kommunikation GmbH, Vienna University of Technology, Energieinstitut der Wirtschaft GmbH, and Austrian Institute of Technology GmbH, *Smart City Wien: Vision 2050, Roadmap for 2020 and Beyond, Action Plan for 2012–15* (2013) <https://www.wien.gv.at/stadtentwicklung/studien/pdf/b008218.pdf> Accessed September 1, 2016.

Climate and Energy Fund, *Smart City Wien* (Vienna: Climate and Energy Fund, 2013) <http://www.smartcities.at/city-projects/smart-cities-en-us/smart-city-wien-en-us/> Accessed September 1, 2016.

CLUE Project Partners, *Practices, Tools and Policies: European Cities Moving Towards Climate Neutrality* (2014) <http://www.clue-project.eu/getfile.ashx?cid=503736&cc=3&refid=18> Accessed March 10, 2016.

A. Cocchia, "Smart and Digital City: A Systematic Literature Review," in Dameri and Rosenthal-Sabroux, *Smart City* (2014) 13–43.

C. Coletta, L. Heaphy, and R. Kitchin, *From the Accidental to Articulated Smart City: The Creation and Work of Smart Dublin* (Maynooth: Maynooth University, 2017) <https://osf.io/preprints/socarxiv/93ga5> Accessed September 12, 2017.

G. Concilio and F. Rizzo, eds, *Human Smart Cities: Rethinking the Interplay Between Design and Planning* (Berlin: Springer, 2016).

M. Cosgrove, W. Harthoorn, J. Hogan, R. Jabbar, M. Kehoe, J. Meegan, and P. Nesbitt, *Smarter Cities Series: Introducing the IBM City Operations and Management Solution* (Armonk, NY: IBM Corporation, 2011) <http://www.redbooks.ibm.com/redpapers/pdfs/redp4734.pdf> Accessed September 14, 2012.

R. Cowley, S. Joss, and Y. Dayot, "The Smart City and Its Publics: Insights from Across Six UK Cities," *Urban Research and Practice* (2017) doi: http://doi.org/10.1080/17535069.2017.1293150.

J.W. Creswell, *Research Design: Qualitative, Quantitative, and Mixed Methods Approaches* (Thousand Oaks, CA: Sage, 2009).

F. Cugurullo, "How to Build a Sandcastle: An Analysis of the Genesis and Development of Masdar City," *Journal of Urban Technology* 20:1 (2013) 23–37.

R.P. Dameri, "Searching for Smart City Definition: A Comprehensive Proposal," *International Journal of Computers and Technology* 11:5 (2013) 2544–2551.

R.P. Dameri, "Comparing Smart and Digital City: Initiatives and Strategies in Amsterdam and Genoa. Are They Digital and/or Smart?" in Dameri and Rosenthal-Sabroux, *Smart City* (2014) 45–88.

R.P. Dameri, *Smart City Implementation: Creating Economic and Public Value in Innovative Urban Systems* (Cham: Spring, 2017).

A. Datta, "New Urban Utopias of Postcolonial India: Entrepreneurial Urbanization in Dholera Smart City, Gujarat," *Dialogues in Human Geography* 5:1 (2015) 3–22.

M. Deakin, ed., *Smart Cities: Governing, Modelling and Analyzing the Transition* (New York City, NY: Routledge, 2014).

M. Deakin and H. Al Wear, "From Intelligent to Smart Cities," *Intelligent Buildings International* 3:3 (2011) 140–152.

M. Deakin and L. Leydesdorff, "The Triple Helix Model of Smart Cities: A Neo-Evolutionary Perspective," in M. Deakin, ed., *Smart Cities: Governing, Modelling and Analyzing the Transition* (New York: Routledge, 2014) 134–149.

Digital City Wien, *Digital City Wien Aktionstag Am 14. September @ Wiener Forschungsfest* (2015) <https://digitalcity.wien/digital-city-wien-wiener-forschungsfest-2015/> Accessed August 24, 2017.

Digital City Wien, *Digitaler Salon* (2016) <https://digitalcity.wien/digitaler-salon/> Accessed August 24, 2017.

L. Dijkstra and H. Poelman, *Cities in Europe: The New OECD-EC Definition* (Brussels: European Commission, 2012) <http://ec.europa.eu/regional_policy/sources/docgener/focus/2012_01_city.pdf> Accessed March 5, 2017.

S. Dirks and M. Keeling, *A Vision of Smarter Cities: How Cities Can Lead the Way Into a Prosperous and Sustainable Future* (Somers, NY: IBM, 2009) <http://www-03.ibm.com/press/attachments/IBV_Smarter_Cities_-_Final.pdf> Accessed February 3, 2012.

S. Dirks, M. Keeling, and J. Dencik, *How Smart Is Your City: Helping Cities Measure Progress* (Somers, NY: IBM Corporation, 2009) <http://public.dhe.ibm.com/common/ssi/ecm/en/gbe03248usen/GBE03248USEN.PDF> Accessed June 6, 2014.

S. Dirks, C. Gurdgiev, and M. Keeling, *Smarter Cities for Smarter Growth: How Cities Can Optimize Their Systems for the Talent-based Economy* (Somers, NY: IBM, 2010) <http://public.dhe.ibm.com/common/ssi/ecm/en/gbe03348usen/GBE03348USEN.PDF> Accessed February 3, 2012.

K.M. Eisenhardt, "Building Theories from Case Study Research," *Academy of Management Review* 14:4 (1989) 532–550.

K.M. Eisenhardt and M.E. Graebner, "Theory Building from Cases: Opportunities and Challenges," *Academy of Management Journal* 50:1 (2007) 25–32.

A. Ersoy, "Smart Cities as a Mechanism Towards a Broader Understanding of Infrastructure Interdependencies," *Regional Studies, Regional Science* 4:1 (2017) 1–6.

H. Etzkowitz and L. Leydesdorff, "The Dynamics of Innovation: From National Systems and "Mode 2" to a Triple Helix of University–industry–government Relations," *Research Policy* 29:2 (2000) 109–123.

European Commission, *Communication from the Commission to the European Parliament, the Council, the European Economic and Social Committee and the Committee of the Regions. Investing in the Development of Low Carbon Technologies (SET-Plan)* (Brussels: European Commission, 2009) <http://eur-lex.europa.eu/legal-content/IT/TXT/PDF/?uri=CELEX:52009DC0519&from=EN> Accessed February 2, 2014.

European Commission, *Call FP7–ENERGY–SMARTCITIES–2012* (European Commission, 2011) <https://ec.europa.eu/research/participants/portal/doc/call/fp7/fp7-energy-smartcities-2012/31559-fiche_fp7-energy-2012-smartcities_en.pdf> Accessed February 10, 2016.

European Commission, *Call FP7-SMARTCITIES-2013* (European Commission, 2012a) <https://ec.europa.eu/research/participants/portal/doc/call/fp7/fp7-smartcities-2013/32801-call_fiche_fp7-smartcities-2013_en.pdf> Accessed February 10, 2016.

European Commission, *Communication from the Commission: Smart Cities and Communities – European Innovation Partnership* (Brussels: European Commission, 2012b) <http://eur-lex.europa.eu/legal-content/IT/TXT/PDF/?uri=CELEX:52009DC0519&from=EN> Accessed February 2, 2014.

European Commission, *Barcelona Is "iCapital" of Europe* (Brussels: European Commission, 2014a) <http://europa.eu/rapid/press-release_IP-14-239_en.htm?locale=en> Accessed March 13, 2014.

European Commission, *Promoting Civil Society Participation in Policy and Budget Processes* (Luxembourg: Publications Office of the European Union, 2014b) <https://europa.eu/capacity4dev/file/26280/download?token=hKjUXKEa> Accessed April 26, 2017.

European Commission, *Horizon 2020 Work Program 2016–2017: Cross-cutting Activities (Focus Areas)* (European Commission, 2016) <http://ec.europa.eu/research/participants/data/ref/h2020/wp/2016_2017/main/h2020-wp1617-focus_en.pdf> Accessed January 20, 2017.

European Innovation Partnership on Smart Cities and Communities, *European Innovation Partnership on Smart Cities and Communities Strategic Implementation Plan* (European Commission, 2013) <http://ec.europa.eu/eip/smartcities/files/sip_final_en.pdf> Accessed March 28, 2017.

J. Exner, "Smart Cities: Field of Application for Planning Support Systems in the 21st Century?," in *Proceedings Computers in Urban Planning and Urban Management 2015* (Cambridge, MA: MIT Press, 2015).

J. Ferrer, "Barcelona's Smart City Vision: An Opportunity for Transformation," *Field Actions Science Reports* Special Issue 16 (2017) 70–75.

K.J. Fietkiewicz and W.G. Stock, "How Smart are Japanese Cities? An Empirical Investigation of Infrastructures and Governmental Programs in Tokyo, Yokohama, Osaka and Kyoto," in T.X. Bui and R.H. Sprague, eds, *Proceedings of the 48th Hawaii International Conference on System Sciences (HICSS)* (Piscataway, NJ: Institute of Electrical and Electronics Engineers, 2015) 2345–2354.

Food and Agriculture Organization, *FAO Strategy for Partnerships with Civil Society Organizations* (Rome: Food and Agriculture Organization, 2013) <http://www.fao.org/3/a-i3443e.pdf> Accessed May 27, 2017.

N. Gardner and L. Hespanhol, "SMLXL: Scaling the Smart City, From Metropolis to Individual," *City, Culture and Society* 12 (2018) 54–61.

A.L. George and A. Bennett, *Case Studies and Theory Development in the Social Sciences* (Cambridge, MA: MIT Press, 2005).

J. Gerring, "What Is a Case Study and What Is It Good For?" *American Political Science Review* 98:2 (2004) 341–354.

M. Gibbert, W. Ruigrok, and B. Wicki, "What Passes As a Rigorous Case Study?" *Strategic Management Journal* 29:13 (2008) 1465–1474.

G.R. Gibbs, *Analyzing Qualitative Data* (Thousand Oaks, CA: Sage, 2007).

R. Giffinger, C. Ferter, H. Kramar, R. Kalasek, N. Pichler-Milanović, and E. Meijers, *Smart Cities: Ranking of European Medium-sized Cities* (Vienna: Vienna University of Technology Centre of Regional Science [SRF], 2007) <http://www.smart-cities.eu/download/smart_cities_final_report.pdf> Accessed May 9, 2012.

D. Gooch, A. Wolff, G. Kortuem, and R. Brown, "Reimagining the Role of Citizens in Smart City Projects," in *UbiComp/ISWC'15 Adjunct: Adjunct Proceedings of the 2015 ACM International Joint Conference on Pervasive and Ubiquitous Computing and Proceedings of the 2015 ACM International Symposium on Wearable Computers* (New York: ACM, 2015) 1587–1594.

G. Grossi and D. Pianezzi, "Smart Cities: Utopia or Neoliberal Ideology?" *Cities: The International Journal of Urban Policy and Planning* 69 (2017) 79–85.

K. Gupta and R.P. Hall, "The Indian Perspective of Smart Cities," in *Proceedings of the 2017 Smart City Symposium Prague (SCSP)* (Piscataway, NJ: Institute of Electrical and Electronics Engineers, 2017).

C. Harrison, B. Eckman, R. Hamilton, P. Hartswick, J. Kalagnanam, J. Paraszczak, and P. Williams, "Foundations for Smarter Cities," *IBM Journal of Research and Development* 54:4 (2010) 1–16.

C. Harrison, J. Paraszczak, and R.P. Williams, "Preface: Smarter Cities," *IBM Journal of Research and Development* 55:1-2 (2011) 1–5.

S. Hartmann, P. Hlava, L. Tiede, M. Kintisch, U. Mollay, C. Schremmer, T. Brajovic, A. Breitfuss, S. Leitner, T. Brus, K. Weninger, and R. Kalasek, *Transformation Agenda Vienna* (2015) <http://urbantransform.eu/wp-content/uploads/sites/2/2015/07/D2.2_Transformation-Agenda-Vienna.pdf> Accessed May 12, 2016.

S. Heissenberger and T. Schuhböck, *Partizipationsprozess Digitale Agenda Wien* (Vienna: City of Vienna, 2015) <http://www.egovernment-wettbewerb.de/praesentationen/2015/stadt_wien_praesentation.pdf> Accessed August 30, 2016.

H. Hemis, W. Schmid, U. Gigler, G. den Boogert, S. Muller, H. Stock, D. Uuong, S. Emery, E. Meskel, P. Weber, J. Jaeger, L. Ljungqvist, S. Geier, A. Olszak, M. Wróblewski, M. Santman, S. Malnar Neralić, N. Mornar, M. Matasović, and M. Zidar, *Integrating Energy in Urban Planning Processes: Insights From Amsterdam/Zaanstad, Berlin, Paris, Stockholm, Vienna, Warsaw and Zagreb* (2017) <http://www.urbanlearning.eu/fileadmin/user_upload/documents/D4-2_Synthesis-report_upgraded_processes_final_170807.pdf> Accessed September 5, 2017.

D. Hemment and A. Townsend, eds, *Smart Citizens* (Manchester: FutureEverything, 2013).

K. Hofstetter and A. Vogl, "Smart City Wien: Vienna's Stepping Stone into the European Future of Technology and Climate," in M. Schrenk, V.V. Popovich, and P. Zeile, eds, *REAL CORP 2011. Change for Stability: Lifecycles of Cities and Regions. The Role and Possibilities of Foresighted Planning in Transformation Processes. Proceedings of 16th International Conference on Urban Planning, Regional Development and Information Society* (Schwechat: Competence Center of Urban and Regional Planning [CORP], 2011) 1373–1382.

R.G. Hollands, "Will the Real Smart City Please Stand Up?," *City: Analysis of Urban Trends, Culture, Theory, Policy, Action* 12:3 (2008) 303–320.

R.G. Hollands, "Critical Interventions into the Corporate Smart City," *Cambridge Journal of Regions, Economy and Society* 8:1 (2015) 61–77.

R.G. Hollands, "Beyond the Corporate Smart City? Glimpses of Other Possibilities of Smartness," in S. Marvin, A. Luque-Ayala, and C. McFarlane, eds, *Smart Urbanism: Utopian Vision or False Dawn?* (New York: Routledge, 2016) 168–184.

I amsterdam, *Amsterdam Smart City Wins City Star Award* (Amsterdam: I amsterdam, 2011) <http://www.iamsterdam.com/city%20star%20award> Accessed June 23, 2015.

IBM Corporation, *IBM Smarter Cities Challenge* (Armonk, NY: IBM, 2017a) <https://www.smartercitieschallenge.org> Accessed March 20, 2017.

IBM Corporation, *IBM Smarter Planet* (Armonk, NY: IBM, 2017b) <http://www-03.ibm.com/ibm/history/ibm100/us/en/icons/smarterplanet> Accessed March 20, 2017.

IDC, *IDC Smart Cities Österreich 2016 Studie* (IDC, 2016) <http://idc-austria.at/de/research/local-studies> Accessed August 10, 2017.

J.S. Katz and J. Ruano, *Smarter Cities Series: Understanding the IBM Approach to Energy Innovation* (Armonk, NY: IBM, 2011) <http://www.redbooks.ibm.com/redpapers/pdfs/redp4739.pdf> Accessed September 14, 2012.

M. Kehoe, M. Cosgrove, S. De Gennaro, C. Harrison, W. Harthoorn, J. Hogan, J. Meegan, P. Nesbitt, and C. Peters, *Smarter Cities Series: A Foundation for Understanding IBM Smarter Cities* (Armonk, NY: IBM, 2011) <http://www.redbooks.ibm.com/redpapers/pdfs/redp4733.pdf> Accessed September 14, 2012.

R. Kitchin, "The Real-time City? Big Data and Smart Urbanism," *GeoJournal* 79:1 (2014) 1–14.

M. Kohno, Y. Masuyama, N. Kato, and A. Tobe, "Hitachi's Smart City Solutions for New Era of Urban Development," *Hitachi Review* 60:2 (2011) 79–88.

N. Komninos, "Intelligent Cities: Variable Geometries of Spatial Intelligence," *Intelligent Building International* 3:3 (2011) 172–188.

N. Komninos, *The Age of Intelligent Cities: Smart Environments and Innovation-for-all Strategies* (New York: Routledge, 2014).

N. Komninos and L. Mora, "Exploring the Big Picture of Smart City Research," *Scienze Regionali: Italian Journal of Regional Science* 1 (2018) 15–38.

J. Kotkin, *The World's Smartest Cities* (Forbes, 2009) <http://www.forbes.com/2009/12/03/infrastructure-economy-urban-opinions-columnists-smart-cities-09-joel-kotkin.html> Accessed September 5, 2016.

K. Kourtit, M. Deakin, A. Caragliu, C. Del Bo, P. Nijkamp, P. Lombardi, and S. Giordano, "An Advanced Triple Helix Network Framework for Smart Cities Performance," in Deakin, *Smart Cities* (2014) 196–216.

T. Kurebayashi, Y. Masuyama, K. Morita, N. Taniguchi, and F. Mizuki, "Global Initiatives for Smart Urban Development," *Hitachi Review* 60:2 (2011) 89–93.

J. Lee and M.G. Hancock, *Toward a Framework for Smart Cities: A Comparison of Seoul, San Francisco and Amsterdam* (Yonsei University and Stanford University, 2012) <http://iis-db.stanford.edu/evnts/7239/Jung_Hoon_Lee_final.pdf> Accessed June 12, 2014.

J. Lee, M.G. Hancock, and M. Hu, "Towards an Effective Framework for Building Smart Cities: Lessons from Seoul and San Francisco," *Technological Forecasting and Social Change* 89 (2014) 80–99.

L. Leydesdorff and M. Deakin, "The Triple-helix Model of Smart Cities: A Neo-evolutionary Perspective," *Journal of Urban Technology* 18:2 (2011) 53–63.

X. Li, H. Yang, W. Li, and Z. Chen, "Public–Private Partnership in Residential Brownfield Redevelopment: Case Studies of Pittsburg," *Procedia Engineering* 145 (2016) 1534–1540.

C. Manville, G. Cochrane, J. Cave, J. Millard, J.K. Pederson, R.K. Thaarup, A. Liebe, M. Wissner, R. Massink, and B. Kotterink, *Mapping Smart City in the EU* (Brussels: European Parliament Directorate-General for Internal Policies, 2014) <http://www.europarl.europa.eu/RegData/etudes/etudes/join/2014/507480/IPOL-ITRE_ET(2014)507480_EN.pdf> Accessed February 5, 2014.

H. March, "The Smart City and Other ICT-led Techno-imaginaries: Any Room for Dialogue with Degrowth?" *Journal of Cleaner Production* (2016) doi: https://doi.org/10.1016/j.jclepro.2016.09.154.

C. Marguerite, N. Pardo Garcia, E. Haslinger, I. Monteverdi, G. Santicelli, and R. Abdurafikov, *CityOpt: Holistic Simulation and Optimization of Energy Systems in Smart Cities: Vienna Demonstrator* (2016) <http://cityopt.eu/Deliverables/D33.pdf> Accessed February 1, 2017.

D. McNeill, "IBM and the Visual Formation of Smart Cities," in Marvin, Luque-Ayala, and McFarlane, *Smart Urbanism* (2016) 34–51.

M.B. Miles and M.A. Huberman, *Qualitative Data Analysis: An Expanded Sourcebook* (Thousand Oaks, CA: Sage, 1994).

L. Mora and R. Bolici, "The Development Process of Smart City Strategies: The Case of Barcelona," in J. Rajaniemi, ed., *Re-city: Future City: Combining Disciplines* (Tampere: Juvenes, 2016) 155–181.

L. Mora and R. Bolici, "How to Become a Smart City: Learning from Amsterdam," in A. Bisello, D. Vettorato, R. Stephens, and P. Elisei, eds, *Smart and Sustainable Planning for Cities and Regions: Results of SSPCR 2015* (Cham: Springer, 2017) 251–266.

L. Mora, R. Bolici, and M. Deakin, "The First Two Decades of Smart-City Research: A Bibliometric Analysis," *Journal of Urban Technology* 24:1 (2017) 3–27.

L. Mora, M. Deakin, and A. Reid, "Combining Co-Citation Clustering and Text-Based Analysis to Reveal the Main Development Paths of Smart Cities," *Technological Forecasting and Social Change* (2018b) doi: https://doi.org/10.1016/j.techfore.2018.07.019.

L. Mora, M. Deakin, and A. Reid, "Smart City Development Paths: Insights from the First Two Decades of Research," in A. Bisello, D. Vettorato, P. Laconte, and S. Costa, eds, *Smart and Sustainable Planning for Cities and Region: Results of SSPCR 2017* (Cham: Springer, 2018a) 403–427.

L. Mora, M. Deakin, and A. Reid, "Strategic Principles for Smart City Development: A Multiple Case Study Analysis of European Best Practices," *Technological Forecasting and Social Change* (2018c) doi: https://doi.org/10.1016/j.techfore.2018.07.035.

P. Muhlmann, *Smart City Wien: The City for Life* (Vienna: TINA Vienna GmbH, 2017) <https://www.publicconsulting.at/fileadmin/user_upload/media/kpc-consulting/Austrian_CC_Workshop_2017/2._P._Muehlmann_Tina_Vienna_Austrian_CC_WS_2017.pdf> Accessed August 5, 2017.

P. Neirotti, A. De Marco, A.C. Cagliano, G. Mangano, and F. Scorrano, "Current Trends in Smart City Initiatives: Some Stylized Facts," *Cities: The International Journal of Urban Policy and Planning* 38 (2014) 25–36.

V. Niaros, "Introducing a Taxonomy of the "Smart City": Towards a Commons-Oriented Approach?," *tripleC* 14:1 (2016) 51–61.

M. Ornetzeder and H. Rohracher, "Of Solar Collectors, Wind Power, and Car Sharing: Comparing and Understanding Successful Cases of Grassroots Innovations," *Global Environmental Change* 23:5 (2013) 856–867.

S. Paroutis, M. Bennett, and L. Heracleous, "A Strategic View on Smart City Technology: The Case of IBM Smarter Cities During a Recession," *Technological Forecasting and Social Change* 89 (2014) 262–272.

M.Q. Patton, *Qualitative Research and Evaluation Methods* (Thousand Oaks, CA: Sage, 2002).

A. Paul, M. Cleverley, W. Kerr, F. Marzolini, M. Reade, and S. Russo, *Smarter Cities Series: Understanding the IBM Approach to Public Safety* (Armonk, NY: IBM, 2011) <http://www.redbooks.ibm.com/redpapers/pdfs/redp4738.pdf> Accessed September 14, 2012.

A. Pollio, "Technologies of Austerity Urbanism: The "Smart City" Agenda in Italy (2011–2013)," *Urban Geography* 37:4 (2016) 514–534.

C. Ratti and A. Townsend, "The Social Nexus," *Scientific American* (September 2011) 42–48.

R.K. Reddy Kummitha and N. Crutzen, "How Do We Understand Smart Cities? An Evolutionary Perspective," *Cities: The International Journal of Urban Policy and Planning* 67 (2017) 43–52.

E. Reviglio, S. Camerano, A. Carriero, G. Del Bufalo, D. Alterio, M. Calderini, A. De Marco, F.V. Michelucci, P. Neirotti, and F. Scorrano, *Smart City: Development Projects and Financial Instruments* (Rome: Cassa depositi e prestiti, 2013) <http://www.cassaddpp.it/static/upload/mon/monographic-report_smart-city.pdf> Accessed February 26, 2014.

C. Robson, *Real World Research: A Resource for Users of Social Research Methods in Applied Settings* (Hoboken, NJ: Wiley & Sons, 1993).

J. Ruano, T. Chao, P. Hartswick, B. Havers, J. Meegan, S. Wasserkrug, and P. Williams, *Smarter Cities Series: Understanding the IBM Approach to Water Management* (Armonk, NY: IBM Corporation, 2011) <http://www.redbooks.ibm.com/redpapers/pdfs/redp4736.pdf> Accessed September 14, 2012.

S. Sauer, "Do Smart Cities Produce Smart Entrepreneurs?," *Journal of Theoretical and Applied Electronic Commerce Research* 7:3 (2012) 63–73.

S. Schaefer, C. Harrison, N. Lamba, and V. Srikanth, *Smarter Cities Series: Understanding the IBM Approach to Traffic Management* (Armonk, NY: IBM, 2011) <http://www.redbooks.ibm.com/redpapers/pdfs/redp4737.pdf> Accessed September 14, 2012.

H. Schaffers, N. Komninos, M. Pallot, M. Aguas, E. Almirall, T. Bakici, J. Barroca, D. Carter, M. Corriou, J. Fernadez, H. Hielkema, A. Kivilehto, M. Nilsson, A. Oliveira, E. Posio, A. Sällström, R. Santoro, B. Senach, I. Torres, P. Tsarchopoulos, B. Trousse, P. Turkama, and J. Lopez Ventura, *Smart Cities As Innovation Ecosystems Sustained by the Future Internet* (2012) <http://hal.archives-ouvertes.fr/docs/00/76/96/35/PDF/FIREBALL-White-Paper-Final2.pdf> Accessed August 24, 2011.

D. Schuurman, B. Baccarne, L. De Marez, and P. Mechant, "Smart Ideas for Smart Cities: Investigating Crowdsourcing for Generating and Selecting Ideas for ICT Innovation in a City Context," *Journal of Theoretical and Applied Electronic Commerce Research* 7:3 (2012) 49–62.

D. Schuurman, L. De Marez, and P. Ballon, "The Impact of Living Lab Methodology on Open Innovation Contributions and Outcomes," *Technology Innovation Management Review* 6:1 (2016) 7–16.

J. Seawright and J. Gerring, "Case Selection Techniques in Case Study Research: A Menu of Qualitative and Quantitative Options," *Political Research Quarterly* 61:2 (2008) 294–308.

C. Selada, "Smart Cities and the Quadruple Helix Innovation Systems Conceptual Framework: The Case of Portugal," in S. Monteiro and E.G. Carayannis, eds, *The Quadruple Innovation Helix Nexus: A Smart Growth Model, Quantitative Empirical Validation and Operationalization for OECD Countries* (New York: Palgrave, 2017) 211–244.

M. Shakir, "The Selection of Case Studies: Strategies and Their Applications to IS Implementation Cases Studies," *Research Letters in the Information and Mathematical Sciences* 3 (2002) 191–198.

D. Shin, "A Critique of Korean National Information Strategy: Case of National Information Infrastructures," *Government Information Quarterly* 24:3 (2007) 624–645.

D. Shin, "Ubiquitous City: Urban Technologies, Urban Infrastructure and Urban Informatics," *Journal of Information Science* 35:5 (2009) 515–526.

D. Shin and T. Kim, "Large-scale ICT Innovation and Policy," in D.F. Kocaoglu, T.R. Anderson, T.U. Daim, A. Jetter, and C.M. Weber, eds, *PICMET 2010 Proceedings: Technology Management for Global Economic Growth* (Piscataway, NJ: Institute of Electrical and Electronics Engineers [IEEE], 2010) 148–161.

S.T. Shwayri, "A Model Korean Ubiquitous Eco-city? The Politics of Making Songdo," *Journal of Urban Technology* 20:1 (2013) 39–55.

Siemens AG, *Our Future Depends on Intelligent Infrastructures* (Munich: Siemens AG, 2014) <https://www.siemens.com/digitalization/public/pdf/siemens-intelligent-infrastructure.pdf> Accessed March 6, 2017.

O. Soderstrom, T. Paasche, and F. Klauser, "Smart Cities as Corporate Storytelling," *City: Analysis of Urban Trends, Culture, Theory, Policy, Action* 18:3 (2014) 307–320.

R.E. Stake, "The Case Study Method in Social Inquiry," *Educational Researcher* 7:2 (1978) 5–8.

R.E. Stake, *The Art of Case Study Research* (Thousand Oaks, CA: Sage, 1995).

R.E. Stake, "Case Studies," in N.K. Denzin and Y.S. Lincoln, eds, *Strategies of Qualitative Inquiry* (Thousand Oaks, CA: Sage, 1998) 86–109.

A. Strauss and J.M. Corbin, *Basics of Qualitative Research: Grounded Theory Procedures and Techniques* (Thousand Oaks, CA: Sage, 1990).

J. Sujata, S. Saksham, T. Godbole, and Shreya, "Developing Smart Cities: An Integrated Framework," *Procedia Computer Science* 93 (2016) 902–909.

H. Tamai, "Fujitsu's Approach to Smart Cities," *FUJITSU Scientific and Technical Journal* 50:2 (2014) 3–10.

A. Townsend, *Smart Cities: Big Data, Civic Hackers, and the Quest for a New Utopia* (New York: WW Norton, 2013).

United Nations, *Arrangements and Practices for the Interaction of Non-Governmental Organizations in All Activities of the United Nations System* (New York: UN Department of Economic and Social Affairs, 1998) <http://www.un.org/documents/ga/docs/53/plenary/a53-170.htm> Accessed May 15, 2017.

A. Vanolo, "Smartmentality: The Smart City as Disciplinary Strategy," *Urban Studies* 51:5 (2014) 883–898.

P. van Waart, I. Mulder, and C. de Bont, "A Participatory Approach for Envisioning a Smart City," *Social Science Computer Review* 34:6 (2016) 708–723.

W. van Winden and D. van den Buuse, "Smart City Pilot Projects: Exploring the Dimensions and Conditions of Scaling Up," *Journal of Urban Technology* 24:4 (2017) 51–72.

D. Velthausz, *Amsterdam Smart City* (Amsterdam: Amsterdam Smart City, 2011) <http://www.slideshare.net/llisa/amsterdam-smart-city-eng-presentation-2-32011-7131457> Accessed August 2, 2016.

J. Viitanen and R. Kingston, "Smart Cities and Green Growth: Outsourcing Democratic and Environmental Resilience to the Global Technology Sector," *Environment and Planning A* 46:4 (2014) 803–819.

N. Walravens, "Mobile City Applications for Brussels Citizens: Smart City Trends, Challenges and a Reality Check," *Telematics and Informatics* 32:2 (2015) 282–299.

Wien Holding GmbH, *Quality of Life for Vienna* (Vienna: Wien Holding GmbH, 2012) <https://www.wienholding.at/tools/uploads/folderbroschueren/WienHolding-English.pdf> Accessed October 20, 206.

Wiener Modellregion and Climate and Energy Fund, *Statusbericht Der Wiener Modellregion "e-mobility on Demand"* (Wiener Modellregion and Climate and Energy Fund, 2014) <https://www.klimafonds.gv.at/assets/Uploads/Themenprojekte/Modellregionen/e-mobility-on-demand-Wien/201504-Statusberichtemobility-on-demand-Wienfinal.pdf> Accessed October 9, 2015.

E. Woods, D. Alexander, R. Rodriguez Labastida, and R. Watson, *UK Smart Cities Index: Assessment of Strategy and Execution for 10 Cities* (Boulder, CO: Navigant Consulting, 2016) <https://www.navigantresearch.com/wp-assets/brochures/Huawei-Navigant-Research-UK-Smart-Cities-Index-White-Paper-5-18-2016.pdf> Accessed August 31, 2017.

World Economic Forum, *The Future Role of Civil Society* (Geneva: WEF, 2013) <http://www3.weforum.org/docs/WEF_FutureRoleCivilSociety_Report_2013.pdf> Accessed May 1, 2017.

T. Yigitcanlar and S.H. Lee, "Korean Ubiquitous-eco-city: A Smart-sustainable Urban Form or a Branding Hoax?" *Technological Forecasting and Social Change* 89 (2014) 100–114.

T. Yigitcanlar and M. Kamruzzaman, "Does Smart City Policy Lead to Sustainability of Cities?" *Land Use Policy* 73 (2018) 49–58.

R.K. Yin, *Case Study Research: Design and Methods* (Thousand Oaks, CA: Sage, 2009).

R.K. Yin, *Applications of Case Study Research* (Thousand Oaks, CA: Sage, 2012).

Y. Yoshikawa, K. Tada, S. Furuya, and K. Koda, "Actions for Realizing Next-generation Smart Cities," *Hitachi Review* 60:6 (2011) 89–93.

T. Zelt, J. Ibel, and F. Tuncer, *THINK ACT: Smart City, Smart Strategy* (Munich: Roland Berger GmbH, 2017) <https://www.rolandberger.com/publications/publication_pdf/ta_17_008_smart_cities_online.pdf> Accessed August 31, 2017.

S. Zygiaris, "Smart City Reference Model: Assisting Planners to Conceptualize the Building of Smart City Innovation Ecosystems," *Journal of the Knowledge Economy* 4:2 (2013) 217–231.

Appendix A

Table A. Census statistics of the EU States members: cities with a population between 1 and 5 million inhabitants

EU CODE	COUNTRY	N° CITIES	CENSUS YEAR	DATA SOURCE
AT	Austria	1	2013	Statistics Austria (http://www.statistik.at)
BE	Belgium	1	2013	Statistics Belgium (http://www.statistik.at)
BG	Bulgaria	1	2012	National Statistical Institute (http://www.nsi.bg)
HR	Croatia	1	2011	Croatian Bureau of Statistics (http://www.dzs.hr)
CY	Cyprus	0	2011	Statistical Service (http://www.cystat.gov.cy)
CZ	Czech Republic	1	2012	Czech Statistical Office (http://www.czso.cz)
DK	Denmark	1	2013	Ministry of Social Affairs and the Interior (http://www.noegletal.dk)
EE	Estonia	0	2011	Statistics Estonia (http://www.stat.ee)
FI	Finland	1	2013	Population Register Centre (http://vrk.fi)
FR	France	2	2010	National Institute for Statistics and Economic Studies (http://www.insee.fr)
DE	Germany	13	2011	Federal Statistical Office (https://www.destatis.de)
GR	Greece	1	2011	Hellenic Statistical Authority (http://www.statistics.gr)
HU	Hungary	1	2012	Hungarian Central Statistical Office (http://statinfo.ksh.hu)
IE	Ireland	1	2011	Central Statistics Office (http://www.cso.ie)
IT	Italy	6	2013	Ancitel (http://portale.ancitel.it)
LV	Latvia	1	2011	Central Statistical Bureau of Latvia (http://www.csb.gov.lv)
LT	Lithuania	1	2013	Statistics Lithuania (http://www.stat.gov.lt)
LU	Luxembourg	0	2014	National Institute of Statistics and Economic Studies of the Grand Duchy of Luxembourg (http://www.statistiques.public.lu)
MT	Malta	0	2011	National Statistics Office (http://nso.gov.mt)
NL	The Netherlands	3	2014	Statistics Netherlands (http://cbs.nl)
PL	Poland	5	2014	Central Statistical Office of Poland (http://stat.gov.pl)
PT	Portugal	1	2011	National Institute of Statistics (http://www.ine.pt)
RO	Romania	2	2013	National Institute of Statistics (http://www.insse.ro)
SK	Slovakia	1	2011	Statistical Office of the Slovak Republic (http://slovak.statistics.sk)
SI	Slovenia	0	2013	Statistical Office of the Republic of Slovenia (http://stat.sl)
ES	Spain	6	2012	Spanish Statistical Office (http://www.ine.es)
SE	Sweden	2	2012	Statistics Sweden (http://scb.se)
UK	United Kingdom	7	2011	Office for National Statistics (http://ons.gov.uk)

Appendix B

Table B. Vienna's smart city development strategy: activities by category and application domain

A: Community Building; B: Strategic Framework; C: Services and Applications; D. Digital infrastructure; C.01. Energy networks; C.02. Air; C.03. Water; C.04. Waste; C.05. Mobility and transport; C.06. Buildings and districts; C.07. Health and Social Inclusion; C.08. Cultural heritage; C.09. Education; C.10. Public safety and security; C.11. E-government; C.12. Other.

ID Code	Name	Category				Application domain											
		A	B	C	D	C.01	C.02	C.03	C.04	C.05	C.06	C.07	C.08	C.09	C.10	C.11	C.12
ACT.0001	AnachB – smart von A nach B			X						X							
ACT.0002	Aspern.mobil		X														
ACT.0003	Boutiquehotel Stadthalle: Stadthotel mit Null-Energie-Bilanz			X							X						
ACT.0004	CASE (Competencies for A sustainable Socio-Economic development)																
ACT.0005	Citybike Wien	X	X	X		X	X			X							
ACT.0006	CITYOPT																
ACT.0007	CLUE (Climate Neutral Urban Districts in Europe)		X	X		X	X			X	X						
ACT.0008	CO2 neutrale Post																
ACT.0009	Die MA 48 Mist App	X	X	X					X								
ACT.0010	Digital Agenda Wien																
ACT.0011	DigitalCity.Wien																
ACT.0012	E-Mobility on Demand			X			X			X							
ACT.0013	E-Taxis			X						X							
ACT.0014	Energiespar-Bim			X						X							
ACT.0015	EOS – Energie aus Klärschlamm			X		X					X						
ACT.0016	EU-GUGLE: Sustainable renovation models for smarter cities																
ACT.0017	Forschungsprojekt SMART.MONITOR																
ACT.0018	INNOSPIRIT																
ACT.0019	INWAPO																
ACT.0020	IREEN																
ACT.0021	LED-Technik in der öffentlichen Beleuchtung			X		X										X	
ACT.0022	Open Government Data			X													
ACT.0023	Optihubs																
ACT.0024	Photovoltaik-Dachgarten			X		X					X						
ACT.0025	Arrowhead			X							X						
ACT.0026	Programm Klimaaktiv Erneuerbare Wärme	X	X	X													
ACT.0027	Smart City Wien Project	X	X	X													
ACT.0028	PUMAS	X	X	X													
ACT.0029	SeniorTab			X						X		X					
ACT.0030	Skopje Urban Transport																
ACT.0031	SCDA (Smart Cities Demo Aspern)			X		X					X					X	

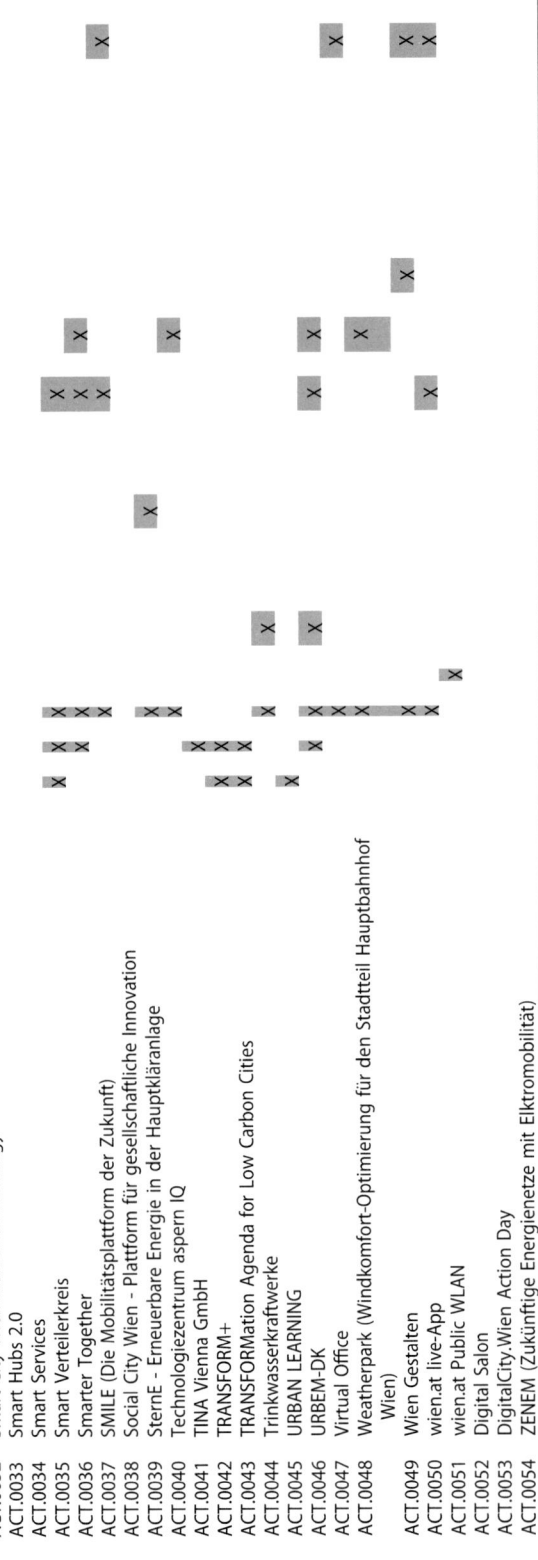

ACT.0032 Smart City Wien Framework Strategy
ACT.0033 Smart Hubs 2.0
ACT.0034 Smart Services
ACT.0035 Smart Verteilerkreis
ACT.0036 Smarter Together
ACT.0037 SMILE (Die Mobilitätsplattform der Zukunft)
ACT.0038 Social City Wien - Plattform für gesellschaftliche Innovation
ACT.0039 SternE - Erneuerbare Energie in der Hauptkläranlage
ACT.0040 Technologiezentrum aspern IQ
ACT.0041 TINA Vienna GmbH
ACT.0042 TRANSFORM+
ACT.0043 TRANSFORMation Agenda for Low Carbon Cities
ACT.0044 Trinkwasserkraftwerke
ACT.0045 URBAN LEARNING
ACT.0046 URBEM-DK
ACT.0047 Virtual Office
ACT.0048 Weatherpark (Windkomfort-Optimierung für den Stadtteil Hauptbahnhof Wien)
ACT.0049 Wien Gestalten
ACT.0050 wien.at live-App
ACT.0051 wien.at Public WLAN
ACT.0052 Digital Salon
ACT.0053 DigitalCity.Wien Action Day
ACT.0054 ZENEM (Zukünftige Energienetze mit Elktromobilität)

Appendix C

Table C. Vienna's smart city development strategy: collaborative ecosystem
RES: Research; IND: Industry; GOV: Government; CIV: Civil Society; OTH: Other.

	Organization		Location		N° of
ID Code	Name	Type	City	Country	Activities
ORG.0001	City of Vienna	GOV	Wien	Austria	24
ORG.0002	TINA Vienna GmbH	GOV	Wien	Austria	15
ORG.0003	AIT Austrian Institute of Technology	RES	Wien	Austria	11
ORG.0004	Vienna University of Technology	RES	Wien	Austria	10
ORG.0005	Neue Urbane Mobilitat Wien GmbH	GOV	Wien	Austria	9
ORG.0006	Siemens Aktiengesellschaft Oesterreich	IND	Wien	Austria	5
ORG.0007	Wien Energie Stromnetz GmbH	GOV	Wien	Austria	5
ORG.0008	VTT Technical Research Centre of Finland	RES	Espoo	Finland	4
ORG.0009	Wien 3420 Aspern Development GmbH	IND	Wien	Austria	4
ORG.0010	Wiener Linien GmbH	GOV	Wien	Austria	4
ORG.0011	Wiener Netze Gmbh	GOV	Wien	Austria	4
ORG.0012	Fachhochschule des BFI Wien	RES	Wien	Austria	3
ORG.0013	City of Amsterdam	GOV	Amsterdam	The Netherlands	3
ORG.0014	Osterreichisches Institut fur Raumplanung (OIR)	RES	Wien	Austria	3
ORG.0015	University of Natural Resources and Life Sciences, Vienna	RES	Wien	Austria	3
ORG.0016	Vienna Business Agency	IND	Wien	Austria	3
ORG.0017	Acciona Infraestructuras	IND	Alcobendas	Spain	2
ORG.0018	Centre Scientifique et Technique du Batiment	RES	Champs Sur Marne	France	2
ORG.0019	City of Hamburg	GOV	Hamburg	Germany	2
ORG.0020	City of Stockholm	GOV	Stockholm	Sweden	2
ORG.0021	City of Turin	GOV	Turin	Italy	2
ORG.0022	Delft University of Technology	RES	Delft	The Netherlands	2
ORG.0023	Ebswien Hauptklaranlage Ges.m.b.H.	IND	Wien	Austria	2
ORG.0024	ETA Umweltmanagement und Technologiebewertung GmbH	IND	Wien	Austria	2
ORG.0025	Fachhochschule Burgenland GmbH	RES	Eisenstadt	Austria	2
ORG.0026	Graz University of Technology	RES	Graz	Austria	2
ORG.0027	Nast Consulting GmbH	IND	Wien	Austria	2
ORG.0028	Osterreichische Post AG	IND	Wien	Austria	2
ORG.0029	Osterreichisches Forschungs- und Prufzentrum Arsenal GesmbH	RES	Wien	Austria	2
ORG.0030	Salzburger Institut für Raumordnung	RES	Salzburg	Austria	2
ORG.0031	Verein DigitalCity.Wien	CIV	Wien	Austria	2
ORG.0032	Vienna University of Economics and Business	RES	Wien	Austria	2
ORG.0033	Wiener Hafen Management GmbH	IND	Wien	Austria	2
ORG.0034	3E	IND	Brussels	Belgium	1
ORG.0035	Aalborg University	RES	Aalborg	Denmark	1
ORG.0036	Abelko Innovation	IND	Lulea	Sweden	1
ORG.0037	Accenture BV	IND	Amsterdam	The Netherlands	1
ORG.0038	AfB Social & Green IT	CIV	Wien	Austria	1
ORG.0039	Agence Parisienne du Climat	CIV	Paris	France	1
ORG.0040	Agenzia Regionale per l'Energia della Liguria	GOV	Genoa	Italy	1
ORG.0041	Airbus Operations	IND	Toulouse	France	1
ORG.0042	Aitia International Informatikai Zartkoruen Mukodo Rt	IND	Budapest	Hungary	1
ORG.0043	Akhela	IND	Cagliari	Italy	1
ORG.0044	Aktiebolaget Skf	IND	Goteborg	Sweden	1
ORG.0045	Alukonigstahl GmbH	IND	Wien	Austria	1
ORG.0046	APC Business Services GmbH	IND	Wien	Austria	1
ORG.0047	Artelys	IND	Paris	France	1
ORG.0048	ASFINAG Service GmbH	IND	Ansfelden	Austria	1

(Continued)

Table C. Continued

Organization		Location			N° of
ID Code	Name	Type	City	Country	Activities
ORG.0049	Aspern Smart City Research GmbH & Co KG	IND	Wien	Austria	1
ORG.0050	ATB-BECKER Photovoltaik GmbH	IND	Absam	Austria	1
ORG.0051	Atos Spain	IND	Madrid	Spain	1
ORG.0052	Austrian Research Promotion Agency	GOV	Wien	Austria	1
ORG.0053	Autorità Portuale di Trieste	GOV	Trieste	Italy	1
ORG.0054	Autorita Portuale di Venezia	GOV	Venice	Italy	1
ORG.0055	Avl List Gmbh	IND	Graz	Austria	1
ORG.0056	Axians ICT Austria GmbH	IND	Wien	Austria	1
ORG.0057	Azienda Lombarda Edilizia Residenziale Milano (ALER)	GOV	Milan	Italy	1
ORG.0058	Bad Radkersburg Beteiligungsgesellschaft mbH	GOV	Bad Radkersburg	Austria	1
ORG.0059	Barcelona Regional, Agencia Desenvolupament Urba	GOV	Barcelona	Spain	1
ORG.0060	Berliner Energieagentur	IND	Berlin	Germany	1
ORG.0061	Bitron	IND	Turin	Italy	1
ORG.0062	Bnearit	IND	Lulea	Sweden	1
ORG.0063	Boliden Mineral	IND	Skelleftea	Sweden	1
ORG.0064	Boutiquehotel Stadthalle	IND	Wien	Austria	1
ORG.0065	Buwog GmbH	IND	Innsbruck	Austria	1
ORG.0066	C2 Smartlight	IND	Jyvaskyla	Finland	1
ORG.0067	CA Technologies	IND	Wien	Austria	1
ORG.0068	CEIT Alanova GmbH	RES	Schwechat	Austria	1
ORG.0069	Cener-Ciemat	RES	Egues	Spain	1
ORG.0070	Centro Ricerche Fiat	RES	Orbassano	Italy	1
ORG.0071	CES Clean Energy Solutions GmbH	IND	Wien	Austria	1
ORG.0072	City of Bratislava	GOV	Bratislava	Slovak Republic	1
ORG.0073	City of Copenhagen	GOV	Copenhagen	Denmark	1
ORG.0074	City of Genoa	GOV	Genoa	Italy	1
ORG.0075	City of Helsinki	GOV	Helsinki	Finland	1
ORG.0076	City of Milan	GOV	Milan	Italy	1
ORG.0077	City of Muenchen	GOV	Munich	Germany	1
ORG.0078	City of Paggaio	GOV	Paggaio	Greece	1
ORG.0079	City of Rome	GOV	Roma	Italy	1
ORG.0080	City of Tampere	GOV	Tampere	Finland	1
ORG.0081	City of Venice	GOV	Venice	Italy	1
ORG.0082	Communaute Urbaine Lyon	GOV	Lyon	France	1
ORG.0083	Commune de Lyon	GOV	Lyon	France	1
ORG.0084	Core As	IND	Herlev	Denmark	1
ORG.0085	Create.at	IND	Wien	Austria	1
ORG.0086	Czech Technical University in Prague	RES	Prague	Czech Republic	1
ORG.0087	D'appolonia	IND	Genoa	Italy	1
ORG.0088	Denkstatt GmbH	IND	Wien	Austria	1
ORG.0089	Dialog Plus	IND	Wien	Austria	1
ORG.0090	Dong Energy Wind Power Holding	IND	Fredericia	Denmark	1
ORG.0091	Edinburgh Napier University	RES	Edinburgh	United Kingdom	1
ORG.0092	EINE	CIV	Padova	Italy	1
ORG.0093	Eistec	IND	Lulea	Sweden	1
ORG.0094	Ekocentrum	IND	Goteborg	Sweden	1
ORG.0095	Electricite de France	IND	Paris	France	1
ORG.0096	Electricite Reseau Distribution France	IND	Paris	France	1
ORG.0097	Enel Distribuzione	IND	Roma	Italy	1
ORG.0098	Energie Tirol	CIV	Innsbruck	Austria	1
ORG.0099	Energie- und Gebaudemanagement GmbH	IND	Wien	Austria	1
ORG.0100	Energie- und Umweltagentur Niederosterreich	CIV	Sankt Polten	Austria	1
ORG.0101	Energieinstitut Vorarlberg	CIV	Dornbirn	Austria	1
ORG.0102	Energy Institute	RES	Zagreb	Croatia	1
ORG.0103	Ente Vasco de la Energia	GOV	Bilbao	Spain	1

(Continued)

Table C. Continued

ID Code	Name	Type	City	Country	N° of Activities
	Organization		Location		N° of
ORG.0104	Ertex Solartechnik GmbH	IND	Amstetten	Austria	1
ORG.0105	Europcar Osterreich	IND	Wien	Austria	1
ORG.0106	Eurotech	IND	Amaro	Italy	1
ORG.0107	Evolaris Next Level Gmbh	IND	Graz	Austria	1
ORG.0108	Evopro Innovation	IND	Budapest	Hungary	1
ORG.0109	Experientia Srl	IND	Turin	Italy	1
ORG.0110	Fagor Electrodomesticos	IND	Mondragon	Spain	1
ORG.0111	Federation of Austrian Industries (IV)	IND	Wien	Austria	1
ORG.0112	Fluidhouse	IND	Jyvaskyla	Finland	1
ORG.0113	Fluidtime Data Services GmbH	IND	Wien	Austria	1
ORG.0114	Fomento de San Sebastian	GOV	Donostia-San Sebastian	Spain	1
ORG.0115	Ford Forschungszentrum Aachen Gmbh	IND	Aachen	Germany	1
ORG.0116	Fotonic i Norden	IND	Skelleftea	Sweden	1
ORG.0117	Free University of Bozen	RES	Bozen	Italy	1
ORG.0118	Freeport of Budapest Logistics Ltd	IND	Budapest	Hungary	1
ORG.0119	French Alternative Energies and Atomic Energy Commission	RES	Paris	France	1
ORG.0120	Fricke Holding GmbH	IND	Heeslingen	Germany	1
ORG.0121	Fully Distributed Systems Ltd	IND	Loughborough	United Kingdom	1
ORG.0122	Gaziantep Buyuksehir Belediyesi	GOV	Gaziantep	Turkey	1
ORG.0123	Gemeente Zaanstad	GOV	Zaandam	The Netherlands	1
ORG.0124	General Directorate of Water Management	GOV	Budapest	Hungary	1
ORG.0125	Gewiss	IND	Cenate Sotto	Italy	1
ORG.0126	Gewista	IND	Wien	Austria	1
ORG.0127	Gewoge AG	IND	Aachen	Germany	1
ORG.0128	Goi Eskola Politeknikoa - Mondragon Unibertsitatea	RES	Mondragon	Spain	1
ORG.0129	Goteborgs Kommun	GOV	Goteborg	Sweden	1
ORG.0130	Grad Zagreb	GOV	Zagreb	Croatia	1
ORG.0131	Grazer Energieagentur	IND	Graz	Austria	1
ORG.0132	GreenIT Consortium	IND	Amsterdam	The Netherlands	1
ORG.0133	Greenovate! Europe	IND	Brussels	Belgium	1
ORG.0134	Grenoble Institute of Technology	RES	Grenoble	France	1
ORG.0135	Hamburg Energie Gmbh	IND	Hamburg	Germany	1
ORG.0136	Hespul Association	CIV	Lyon	France	1
ORG.0137	HMP Beratungs GmbH	IND	Wien	Austria	1
ORG.0138	Hofor	IND	Copenhagen	Denmark	1
ORG.0139	Honeywell Czech Republic	IND	Prague	Czech Republic	1
ORG.0140	IBA Hamburg GmbH	IND	Hamburg	Germany	1
ORG.0141	iC consulenten ZT GmbH	IND	Wien	Austria	1
ORG.0142	ICT Austria	IND	Wien	Austria	1
ORG.0143	IK4-IKERLAN	RES	Mondragon	Spain	1
ORG.0144	IKARUS Security Software GmbH	IND	Wien	Austria	1
ORG.0145	Indra Sistemas	IND	Alcobendas	Spain	1
ORG.0146	Indra Software Labs SLU	IND	Alcobendas	Spain	1
ORG.0147	Infineon Technologies Austria	IND	Villach	Austria	1
ORG.0148	Instituto de Desenvolvimento de Novas Tecnologias (Uninova)	RES	Caparica	Portugal	1
ORG.0149	Instituto Superior de Engenharia do Porto	RES	Porto	Portugal	1
ORG.0150	Integrasys	IND	Sevilla	Spain	1
ORG.0151	Joseph Fourier University	RES	Grenoble	France	1
ORG.0152	Kaferhaus GmbH	IND	Wien	Austria	1
ORG.0153	Kapraluv mlyn	RES	Brno	Czech Republic	1
ORG.0154	Kelag	IND	Klagenfurt	Austria	1
ORG.0155	Landschaftsplanerin Martina Jauschneg	IND	Wien	Austria	1
ORG.0156	LeasePlan Austria Fleet Management GmbH	IND	Wien	Austria	1
ORG.0157	LKAB	IND	Lulea	Sweden	1
ORG.0158	Lulea University of Technology	RES	Lulea	Sweden	1

(*Continued*)

Table C. Continued

| Organization | | Location | | N° of |
ID Code	Name	Type	City	Country	Activities
ORG.0159	Lyse Energi	IND	Stavanger	Norway	1
ORG.0160	Magillem Design Services	IND	Paris	France	1
ORG.0161	Malopolska Region	GOV	Krakov	Poland	1
ORG.0162	Manchester City Council	GOV	Manchester	United Kingdom	1
ORG.0163	Masaryk University	RES	Brno	Czech Republic	1
ORG.0164	Mazovia Development Agency Plc	IND	Warsaw	Poland	1
ORG.0165	Métropole Nice Cote d'Azur	GOV	Nice	France	1
ORG.0166	Metso Minerals	IND	Helsinki	Finland	1
ORG.0167	Miasto Stoleczne Warszawa	GOV	Warsaw	Poland	1
ORG.0168	Micro Dators Ltd	IND	Riga	Latvia	1
ORG.0169	Midroc Electro	IND	Sandviken	Sweden	1
ORG.0170	Ministry of Transportation of the Czech Republic	GOV	Prague	Czech Republic	1
ORG.0171	MOOSMOAR Energies OG	IND	Oblarn	Austria	1
ORG.0172	Msg systems AG	IND	Ismaning	Germany	1
ORG.0173	Munchner Verkehrs- und Tarifverbund (MVV)	IND	Munich	Germany	1
ORG.0174	Municipality of Nova Gorica	GOV	Nova Gorica	Slovenia	1
ORG.0175	N/A (Undifined Planning Agency in Vienna 1)	OTH	Wien	Austria	1
ORG.0176	N/A (Undifined Planning Agency in Vienna 2)	OTH	Wien	Austria	1
ORG.0177	Neogrid Technologies Aps	IND	Aalborg	Denmark	1
ORG.0178	NetApp GmbH	IND	Sunnyvale	United States	1
ORG.0179	NetHotels AG	IND	Wien	Austria	1
ORG.0180	Noda Intelligent Systems	IND	Karlshamn	Germany	1
ORG.0181	Nord-Trondelag Elektrisitetsverk	IND	Steinkjer	Norway	1
ORG.0182	NTT Data	IND	Milan	Italy	1
ORG.0183	NXP Semiconductors GA Gmbh	IND	Hamburg	Germany	1
ORG.0184	OBB-Holding AG	IND	Wien	Austria	1
ORG.0185	Orona S. Coop	IND	Hernani	Spain	1
ORG.0186	Osterreichisches Forschungs- und Prufzentrum Arsenal Ges.mBH	RES	Wien	Austria	1
ORG.0187	Outokumpu Stainless	IND	Espoo	Finland	1
ORG.0188	Ove Arup & Partners	IND	London	United Kingdom	1
ORG.0189	Pannon Gazdasagi Halozat Egyesulet	CIV	Gyor	Hungary	1
ORG.0190	Personal Space Technologies	IND	Amsterdam	The Netherlands	1
ORG.0191	Politecnico di Milano	RES	Milan	Italy	1
ORG.0192	Politecnico di Torino	RES	Turin	Italy	1
ORG.0193	Port of Koper	IND	Koper	Slovenia	1
ORG.0194	Punkt.wien GmbH	IND	Wien	Austria	1
ORG.0195	Quintessenz	CIV	Wien	Austria	1
ORG.0196	Raiffeisen-Leasing GmbH	IND	Wien	Austria	1
ORG.0197	Raintime Gmbh	IND	Leopoldsdorf	Austria	1
ORG.0198	Regionalni Rozvojova Agentura Usteckeho Kraje	GOV	Usti Nad Labem	Czech Republic	1
ORG.0199	Republic of Slovenia	GOV	Ljubljana	Slovenia	1
ORG.0200	Rhone-Alps Region	GOV	Lyon	France	1
ORG.0201	Riga Technical University	RES	Riga	Latvia	1
ORG.0202	Risorse Per Roma Spa	IND	Roma	Italy	1
ORG.0203	Royal Institute of Technology	RES	Stockholm	Sweden	1
ORG.0204	Samsung Electronics Austria GmbH	IND	Wien	Austria	1
ORG.0205	SAP	IND	Walldorf	Germany	1
ORG.0206	SCHIG mbh	IND	Wien	Austria	1
ORG.0207	Schneider Electric Industries	IND	Rueil Malmaison	France	1
ORG.0208	Seluxit ApS	IND	Aalborg	Denmark	1
ORG.0209	SeniorPad GmbH	IND	Wien	Austria	1
ORG.0210	SERA energy & resources e.U.	IND	Wien	Austria	1
ORG.0211	Siemens AG	IND	Munich	Germany	1
ORG.0212	Sirris	IND	Brussels	Belgium	1
ORG.0213	Slovenska plavba a prístavy AS	IND	Bratislava	Slovak Republic	1
ORG.0214	Slovenska rada pre zelené budovy	IND	Bratislava	Slovak Republic	1

(Continued)

Table C. Continued

	Organization		Location		N° of
ID Code	Name	Type	City	Country	Activities
ORG.0215	Social City Wien	CIV	Wien	Austria	1
ORG.0216	SPAR Osterreichische Warenhandels AG	IND	Wien	Austria	1
ORG.0217	Sprinte	IND	Saint Etienne	France	1
ORG.0218	Stadt Aachen	GOV	Aachen	Germany	1
ORG.0219	STAWAG Stadtwerke Aachen AG	IND	Aachen	Germany	1
ORG.0220	Stiftelsensintef	RES	Trondheim	Norway	1
ORG.0221	Stmicroelectronics	IND	Agrate Brianza	Italy	1
ORG.0222	SYCUBE Informationstechnologie GmbH	IND	Wien	Austria	1
ORG.0223	SynergieKomm	RES	Bonn	Germany	1
ORG.0224	Tampere University of Technology	RES	Tampere	Finland	1
ORG.0225	Taxi 31300 Vermittlungs GmbH	IND	Wien	Austria	1
ORG.0226	TBuilding Testing and Research Institute	RES	Bratislava	Slovak Republic	1
ORG.0227	tbw research GesmbH	IND	Wien	Austria	1
ORG.0228	Technical University of Denmark	RES	Kgs. Lyngby	Denmark	1
ORG.0229	TECNALIA Research & Innovation	RES	Derio	Spain	1
ORG.0230	Tekniker	RES	Eibar Guipuzcoa	Spain	1
ORG.0231	Telekom Austria Group	IND	Wien	Austria	1
ORG.0232	Terra Institute Srl	IND	Bressanone	Italy	1
ORG.0233	Thales Communications & Security	IND	Gennevilliers	France	1
ORG.0234	THT Control	IND	Tampere	Finland	1
ORG.0235	Treberspurg & Partner Architekten Ziviltechniker GmbH	IND	Wien	Austria	1
ORG.0236	Ulma Embedded Solutions	IND	Onati	Spain	1
ORG.0237	Ulrich Walter GmbH	IND	Diepholz	Germany	1
ORG.0238	Università di Vechta	RES	Vechta	Germany	1
ORG.0239	University of Bologna	RES	Bologna	Italy	1
ORG.0240	University of Gothenburg	RES	Goteborg	Sweden	1
ORG.0241	University of Hagen	RES	Hagen	Germany	1
ORG.0242	University of Oulu	RES	Oulu	Finland	1
ORG.0243	University of Warwick	RES	Coventry	United Kingdom	1
ORG.0244	Verejne Pristavy	GOV	Bratislava	Slovak Republic	1
ORG.0245	Verkehrsverbund Ost-Region (VOR)	IND	Wien	Austria	1
ORG.0246	via donau - Austrian Waterway mbH	IND	Wien	Austria	1
ORG.0247	Ville de Paris	GOV	Paris	France	1
ORG.0248	Voestalpine Krems GmbH	IND	Krems An Der Donau	Austria	1
ORG.0249	Vossloh Kiepe GmbH	IND	Dusseldorf	Germany	1
ORG.0250	Wapice Ltd	IND	Vaasa	Finland	1
ORG.0251	Weatherpark GmbH	IND	Wien	Austria	1
ORG.0252	Wiener Wissenschafts-, Forschungs- und Technologiefonds	RES	Wien	Austria	1
ORG.0253	WienIT EDV Dienstleistungs-gesellschaft mbH & Co KG	IND	Wien	Austria	1
ORG.0254	WIPARK Garagen GmbH	IND	Wien	Austria	1
ORG.0255	Zense Technology	IND	Norresundby	Denmark	1

"Mapping" Smart Cities

Becky P. Y. Loo ⓘ and Winnie S. M. Tang

ABSTRACT

Smart cities are designed to use data to optimize resources, maintain sustainability, and improve people's quality of life. While many urban technologies are employed to make cities "smart," one constellation of technologies has been less examined in the academic literature—digital maps and the spatial data infrastructure. This paper is an attempt to systematically review the functions and evolution of digital maps and the spatial data infrastructure, with examples from Asia and beyond, in supporting and making smart cities possible. Based on the conceptual framework and empirical case studies, four major research directions of smart mapping are identified to better support smart city initiatives.

Introduction

Over the last several decades, advances in telecommunications, notably the Internet, combined with the development of electronic devices have transformed people's everyday lives and mind sets (Loo, 2012). The amazing development of telecommunications with rapidly increasing speed and capacity has increased the amount and range of information transmitted on the Internet at exponential rates. At the same time, a wide diversity of electronic devices have been developed that are cheap, compact, and customer-specific. These simple and light devices often collect basic environmental attributes of a place/location by capturing images/videos, measuring temperature, and detecting movements. When these devices are connected to the Internet, a whole series of real-time, spatio-temporal data can be captured. The most ubiquitous of these devices, smartphones, have become more and more sophisticated, combining traditional functions of communications and entertainment with various real-time and online platforms available to users via multifarious applications (apps). They are getting ever-more sophisticated and multi-functional, replacing other single-purpose electronic devices and those with limited functions like personal digital assistants, digital cameras, and digital music players. All these changes have accelerated greatly over the last two decades. While the term the "Internet of Things" (IoT) has been used earlier (Ashton, 2009), its first appearance in the academic literature can be traced to Gershenfeld et al. (2004).

This paper has two objectives and two sets of research questions. The first objective is to systematically review the functions and the evolution of digital maps and their applications in supporting and making smart cities possible. The key research questions are:

- How did digital maps evolve over time?
- With the constant generation of huge volumes of spatio-temporal data from IoT, why and how has digital mapping pushed the boundaries of both conceptual and methodological state-of-the-art data analysis?
- What are the potential benefits of using digital maps to integrate, visualize, and analyze data in a smart city?

The second objective is to identify the major challenges and propose some major research directions for smart mapping to better support various smart cities initiatives. The key research questions are:

- What are challenges in relation to (1) data collection, (2) data analysis, (3) institutional set-up, and (4) users of data?
- What major advancements have been made?
- Where can good examples be found?
- What else needs to be done to realize the full potential of smart mapping as a disruptive, innovative urban technology?

Literature Review and Research Gap

There is no universally accepted definition of a smart city despite the fact that many researchers have attempted to provide a unified definition of the term (Angelidou, 2014; Albino et al., 2015; Ching and Ferreira, 2015; Ramaprasad et al., 2017). For instance, IBM (2009:9) described a smarter city as "one that uses technology to transform its core systems and optimize the return from largely infinite resources." And Cisco (2012:2) defined smart cities as those which adopt "scalable solutions that take advantage of information and communications technology (ICT) to increase efficiencies, reduce costs, and enhance quality of life."

In 2008, Giffinger et al. (2008) described the main characteristics of a smart city. This work was further developed by Cohen (2013), an urban and climate strategist, into the "smart city wheel" with the purpose of tracking the progress of smart cities. The six components/domains were further elaborated as follows:

(1) Smart Economy: Cities should stimulate entrepreneurship, innovation, productivity, and local and global interconnectivity.
(2) Smart Governance: Cities should enable supply and demand side policy, transparency, and open data through ICT and e-government integration.
(3) Smart Environment: Cities should implement sustainability through the use of buildings, green energy, and urban planning.
(4) Smart People: Cities should provide good education, create an inclusive society, and embrace creativity.

(5) Smart Mobility: Cities should prioritize clean and non-motorized options, integrate ICT, and provide mixed-modal access.
(6) Smart Living: Cities should be culturally vibrant and safe, and promote good health.

From a slightly different perspective, Gil-Garcia et al. (2015) identified 10 core components of public services: city administration and management; policies and other institutional arrangements; governance; public engagement and collaboration; human capital and creativity; knowledge economy and pro-business environment; built environment and city infrastructure; natural environment and ecological sustainability; ICT and other technologies; and data and information for analyzing the concept of smart cities. In the paper, they agreed that the 10 identified components were only constructed for analytical purposes and some of them could be combined or could be categorized into more than one component.

Although scholars may use slightly different definitions, it is commonly agreed that the smart city concept is primarily about people and must be people-centric so that citizens can have a safe, efficient, and sustainable living environment and better quality of life. In other words, the concept of smart cities distinguishes itself by serving specific purposes – from resource optimization and maintaining sustainability to improving people's quality of life. With the development of ICT, one of the associated opportunities (as well as challenges) has been the enormous multiplication of data generated from the continuous flow of information (typically at regular intervals of one minute or even less) by different sensors across the city and by smartphones that people use for a wide range of purposes. Given that IoT and related applications are the main urban technologies nowadays, there has been a growing body of research related to the social, as well as technical aspects of these technologies (Kim, 2016; Lee and Lee, 2015). In addition, the big data associated with the rise of IoT has generated renewed interests in the science of cities (Batty et al., 2010). However, there is a lack of a clear framework of utilizing big data to reveal generalizable theoretical patterns or inform decision-makers. Within a city, maps are critical tools for providing a wide range of location data and spatial information such as roads with their speed limits, slopes, and other restrictions; shops with their products or services and user ratings, and other information like building shapes, bus schedules, and transit routes in a user-friendly and intuitive format.

To fill this research gap, this paper explores smart mapping as a means of making the best use of the spatio-temporal data from IoT and other sources. Smart mapping is a term describing the smart use of digital maps and related applications to support all six domains of the smart city wheel. Figure 1 is an attempt to show the conceptual framework. As a disruptive innovative urban technology, smart mapping requires a fundamental rethinking of key issues related to (1) data collection, (2) data analysis, (3) institutional set-up, and (4) users of data (See Figure 1). These four research directions will be discussed below with examples from Asia and beyond.

Methodology

To achieve the first objective and address the first series of research questions, this paper first traces the evolution of digital maps in different parts of the world and examines how smart mapping can support the different domains of a smart city. Then, the second

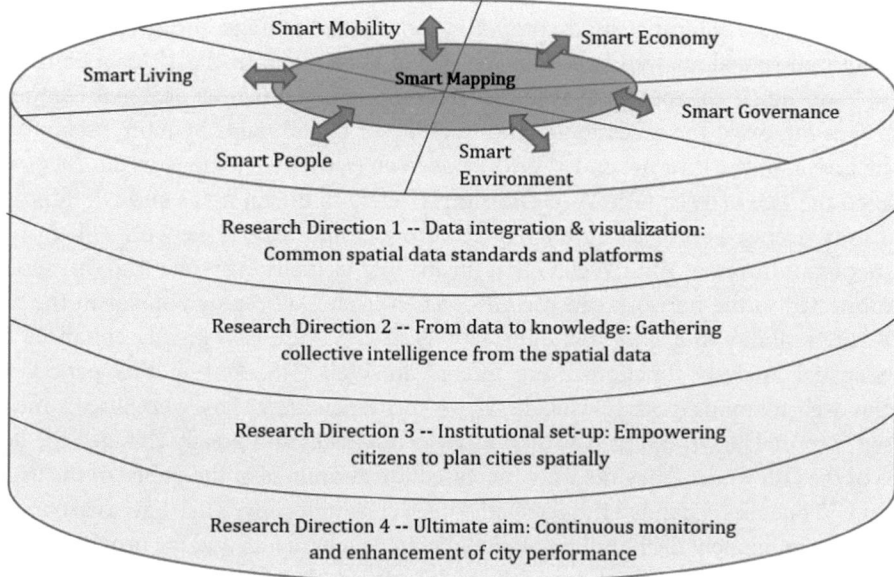

Figure 1. A conceptual framework of "mapping" smart cities

objective and series of research questions are addressed through identifying the major challenges and proposing major research directions for smart mapping to better support various smart cities initiatives. It is recognized that smart mapping will require a fundamental rethinking of key issues related to data collection, data analysis, institutional set-up, and users of data. As this paper aims to go beyond theories to examine planning and development issues of cities, it does not follow the typical research methodology of primary data collection or a meta-analysis of research papers included in scientific databases. The discussion is informed by the researchers' insights into spatial formations of information and communication technologies, and knowledge production practices from their professional and research work. Where good practices are identified, they came from the researchers' experience gained through being directly involved in or communicating with key stakeholders through fieldwork.

Why Smart Mapping?

The world's first computerized GIS, Canada Geographic Information System (CGIS), was developed in the 1960s by Roger Tomlinson (1968), who is generally considered the "father of GIS". CGIS was developed for the Canada Land Inventory to store and support planning analysis of a large amount of land use and natural resources data covering the whole territory of Canada using a layer approach on which modern GIS is based. The digital maps in CGIS were stored in mainframe computers but could only be visualized in hardcopy maps generated from line printers.

With emerging minicomputers and UNIX-based graphics workstations, development of GIS and digital mapping advanced to another stage with a more powerful and accessible computing platform. The world's largest GIS software company, Esri, launched their first version of commercialized GIS software, ArcINFO, on minicomputers in 1982. At that time, digital maps could be visualized dynamically at different scales in large-format graphics monitors. The increased computing power and storage also greatly facilitated computerized spatial analysis based on the digital maps data. In the 1990s, GIS became available on much cheaper Intel-based microcomputers, known as personal computers (PC). This increased the accessibility and usability of digital maps by more users.

The arrival of the Internet and World Wide Web (www) technology in the 2000s may be called the Age of Web GIS. Web GIS simply refers to digital maps and GIS functions that can be accessed and used through a web browser and TCP/IP network either on the Internet or an Intranet. With Web GIS, it means that virtually everyone who can access a PC connected to the network can use GIS. When Web 2.0 became popular in the mid-2000s, the usability and user friendliness of Web GIS were also greatly enhanced and more spatial analysis functions were offered in Web GIS. During this period, two popular web mapping portals, Google Maps and OpenStreetMap, were also launched. These two portals have emerged as disruptive technologies and greatly changed the landscape of the GIS world. They not only increased the awareness of the public of the usefulness of GIS but also expanded the geospatial market significantly. They have two common characteristics, namely the bundling of digital map data with GIS and the provision of web map services and Application Programming Interface (API) for the developers to embed GIS in their own applications.

When the first iPhone was introduced in 2009, we entered the age of the smartphone. With the popularity of smartphones in the 2010s, mobile GIS has also taken off. Nowadays, almost everyone's smartphone has a mapping app so that one can access and use a digital map anywhere, anytime. Riding on the popularity of Web GIS and mature cloud computing technologies, GIS has also moved from the client server model to the cloud-based model. Looking forward, Goodchild (2015) suggests that big data will be the main driver of GIS and digital mapping going forward.

With the emergence of e-societies, IoT are everywhere. With moving or static sensors, mobile devices collecting all kinds of open data in different formats like text, markup language, and image, how to coordinate the use and make these valuable data resources available to society for the sake of developing a smart city is a key issue. It is generally considered that at least 80 percent of data is geospatially referenced nowadays (Hahmann et al., 2011; Robinson, 2008). And it was estimated by the US Office of Management and Budget that 80 to 90 percent of all government information has a spatial component (The White House, 2002). At the minimal level, these data incorporate the locations of the sensors, which may be fixed/mounted or mobile (for example, mounted on a vehicle or a drone). Moreover, all data are "stamped" in both space and time depending on the time interval at which the data are recorded and transmitted back to the cloud servers. Theoretically, all actions taken by people on their smart devices (smartphones and other e-devices) equipped with Global Positioning System (GPS) generate "digital traces"—essentially, all data are "stamped" simultaneously in both space and time. However, unlike fixed sensors, these smart devices are mobile in cities, giving rise to dynamic spatio-temporal data.

While the complexity of handling dynamic temporal data is high already, the addition of the spatial dimension has really pushed the boundaries of both conceptual and methodological state-of-the-art data analysis. The potential benefits of using the spatial dimension to integrate, visualize, and analyze data in a smart city are enormous in all six major

Table 1. Examples of smart mapping applied to the smart city wheel

Domains of the Smart City Wheel	Examples
Smart Economy	Users of smart maps are able to locate local businesses, not only can they save time but they can gain product information from a wider choice of vendors. Smart maps could help local businesses earn $2.2 billion and save shoppers 12 million hours from wasted searches (Dalberg Global Development Advisors, 2015).
Smart Environment	By reducing fuel wastage, Indian cities could save 1 million metric tons of CO_2 annually (Dalberg Global Development Advisors, 2015).
Smart Government	Smart maps could improve urban management by helping detect violations, improving accountability, ensuring that infrastructure and government offices are optimally placed and streamlining planning processes (Dalberg Global Development Advisors, 2015).
Smart Living	With images from live camera feeds, smart maps helped police officers quickly identify and apprehend suspects within minutes (Esri Malaysia, 2015), which helped create a safe living environment for citizens.
Smart Mobility	Users of smart maps could get real-time traffic conditions on the road and select alternative routes to avoid traffic congestion based on smart maps (Weinmann, 2014).
Smart People	Smart maps empower citizens to participate in smart city development and facilitate dialogue between citizens and their industrial and governmental counterparts, ensuring the citizen's voice is included in new solutions (Metz, 2014).

domains. In particular, the local contexts make smart mapping essential when pushing the smart city agenda forward in specific cities (Loo, 2007; Loo and Wong, 2002). To illustrate, depending on the level and nature of economic development of the city, the focus and details of measures to promote a smart economy need to be different. Under a "smart economy," developing "smart finance" should be a priority in a global financial city, while developing "smart tourism" should be considered in a historical tourist city. When smart city measures are implemented, the effects will vary spatially. For instance, policies to reduce harmful air pollutants in a city under "smart environment" need to be targeted at the "hot spots" and to be monitored spatially. Therefore, it is important to push for smart mapping to benefit and speed up the realization of various smart core components simultaneously, and be able to cater to the wide geographical diversity of cities (See Figure 1). Incorporating the spatial dimension in the conceptual and methodological state-of-the-art data analysis has been taking place in different sub-fields like travel behavior analysis (Lam et al., 2014), crime fighting (Esri Malaysia, 2015), logistics and delivery, emergency response, disaster recovery, tourism, transportation, local retail, disease prevention, event management, crowd control, energy generation and consumption forecast, civic engagement and city planning (Dalberg Global Development Advisors, 2015; Rehman et al., 2017; Li et al., 2016; Tascikaraoglu, 2017; Baucells et al., 2016; Celdran-Bernabeu et al., 2018; Zhao et al., 2014; Pfeiffer and Stevens, 2015; Sadiq et al., 2015). Some more examples, grouped by the six domains of the smart city wheel, are given in Table 1.

Four Major Directions of Smart Mapping

While supporting all major domains of smart cities, smart mapping does require a fundamental rethinking of key issues related to data collection, data analysis, institutional set-up, and users of data. To reiterate, there will need to be changes in the ways data are collected, how the data are analyzed, who is responsible to take action, and who can/will use the data (See Figure 1). These changes are related to common spatial data standards and platforms, gathering collective intelligence from the spatial data, empowering citizens to plan cities spatially, and the continuous monitoring and enhancement of city performance. Under each of the four research directions, the challenges created by the exponential growth of real-time socio-temporal data, the directions of change required, some good practices, and implications for policymakers and smart city practice are described.

Setting up of Common Spatial Data Standards and Platforms

With smart mapping, the approach to data collection and the way of using data will have to be changed. Currently, data are often collected by individual government departments under specific projects. These data serve specific purposes and are often owned exclusively by the agency collecting the data. With smart mapping, this needs to be changed. Following our discussion above with the generation of massive amounts of IoT data 24/7 in different domains, a major challenge is to integrate, store, and analyze them in serving meaningful purposes. In relation, there needs to be a Spatial Data Infrastructure (SDI) for the efficient and effective use of geospatial data and information, which underpins analysis and decision-making for environmental, social, and economic growth. The term SDI was first used in 1993 by the US National Research Council to denote "a framework of technologies, policies and institutional arrangements that together facilitate the creation, exchange, and use of geospatial data and related information resources across an information-sharing community" (Esri, 2010:3). In other words, SDI ensures that geospatial data and standards are used to create authoritative datasets and policies that support it. One of the key components of SDI is the open-standard-based API which provides a standardized and relatively simple way for people to integrate geospatial data to build applications.

A number of countries or regions have recognized SDI as an important digital foundation for city development. The United States has a national SDI—the GeoPlatform.gov website where different geographic information has been integrated and shared with the public by APIs. The European Union has also built its own SDI called Infrastructure for Spatial Information in European Community (INSPIRE). INSPIRE is a spatial data infrastructure for the purposes of EU environmental policies and policies or activities which may have an impact on the environment. By adopting common implementing rules, it ensures the geographic datasets and services are understood, compatible, and usable across European countries. Therefore, the effective access and sharing of environmental spatial information among public sector organizations and the public through the INSPIRE Geoportal is possible. Comprehensive spatial datasets and services such as transport networks, species distribution, and farming and mineral resources can be found on the portal.

The development of smart cities hinges on the availability of SDI that leverages the power of open data for professionals and the public (Chow, 2017). With SDI for all stakeholders to exchange and integrate geospatial data, it not only becomes possible for all people to pinpoint more accurately a location in real time, but also facilitates economic development, urban planning, transportation and logistics, car navigation, and many other applications.

The role and importance of SDI could be illustrated by the Reference Model for a Smart City (Deloitte, 2017), which consisted of a data platform layer to aggregate data and information that a smart city collects and manages. In the absence of the data platforms, it would be costly and ineffectively for each of the government departments and business organizations to maintain their own spatial and temporal data for analysis. Applications in all areas of the six components (smart economy, smart governance, smart environment, smart people, smart mobility, and smart living) would become difficult to develop without the appropriate data analytics.

Beyond the multi-nation-regional and national levels, sub-national and city-level SDIs are emerging due to the digital transformation of city development and operations. In 2016, the Los Angeles city council launched *GeoHub*, an online portal providing location-based data. It contains more than 500 categories of real-time (or near real-time) information obtained from various government departments, such as traffic black-spots, temporary road closures, and accidents, and allows anyone to access live, continuously updated data directly from the city's database, rather than as a static download. This enables government agencies, public and private organizations, and application developers to access the latest information. Because of the platform, the city was ranked the highest in open data application in the United States for the year. By opening data to everyone, Eric Garcetti, the Mayor of the city, hopes to "make city operations more efficient, stimulate partnerships between the city and the community, and give residents a greater controlling stake in government." It can also boost the creativity of local people to relieve the problems created by economic downturns and climate change. The key to the success of this vision is the collaboration between different government departments, civil organizations, and systems to ensure interoperability. According to Lilian Coral, the chief data officer of Los Angeles City, her team has lobbied more than 60 government departments to share spatial data. She pointed out that to promote collaboration between multiple parties, the portal has to have a standard of data format; and its API should be all-embracing so that the data can be downloaded in various formats like Keyhole Makeup Language (KML) and Shapefile (.shp). Promoting app development is one of the main objectives of *GeoHub*. In fact, it has spawned many already, including the *Clean Streets Index* (shows street cleanliness), *Street Wize* (displays current and upcoming road works), *Vision Zero Los Angeles* (shows traffic accident black spots), etc. These apps have good reputations and are widely used by the public, creating a sense of accomplishment for the developers. As a result, more organizations are willing to open data and even participate in the design of apps. There are over 600 kinds of datasets on the platform now.

In Asia, the Singapore Geospatial Collaborative Environment (SG-SPACE) program provides a cross-agencies data-sharing platform, which can facilitate better policy, decision-making, governance, and day-to-day operation. It is also targeted to be extended to private sectors for value and knowledge creation. According to Loh and Khoo (2010:111),

> The most important issue is whether the geospatial data are indeed interoperable and ready for integration. To ensure sustainability, we need to create an eco-system and an environment where a strong framework is in place.

In Hong Kong, the Development Bureau of HKSAR Government is promoting the establishment of Common Spatial Data Infrastructure (CSDI). CSDI is a professional, comprehensive, large-scale digital spatial infrastructure which is an important foundation for a smart city. It can support various apps and services. A CSDI allows government departments and public and private organizations to consolidate and exchange various data, such as road networks, plots, land use, underground pipelines, urban planning require-ments, real-time traffic conditions, and local weather conditions. The platform aims to promote the ability not only of government departments and public organizations, but also the ability of citizens, to develop broader functional apps based on spatial data. Tra-ditionally, HKSAR Government agencies like the Departments of Lands, Highways, Civil Engineering, and Development have used dedicated spatial data. However, sharing such information with each other or making it available to the public was not yet a common practice. Nevertheless, there are encouraging initiatives like the GeoInfo Map developed by the Lands Department that is open for public use and contains over 180 kinds of spatial data provided by 26 government departments. The Planning Department has also launched a Statutory Planning Portal, which allows citizens to search for planning and zoning information. It has recorded over 16 million page views in a single month. Moreover, the Government has offered over 6,000 data sets in 18 categories through the data.gov.hk website since March 2015. The local technology sector, however, has cri-ticized the government for providing most of the public information in Excel, CSV, or PDF format instead of the API format which can be directly used in program or app devel-opment. This forms an obstacle for public use. Furthermore, the information is not updated as frequently as it could be. In Hong Kong, the API was developed for use in the platform DATA.GOV.HK to provide software and application developers with differ-ent perspectives and means of using the original datasets shared by public agencies. In the initial stage, it is expected that government agencies should take the initiative to set up the SDI with API which can take up various data formats to be used among government departments, businesses, and the public. Critics argued that an open API could create new customer markets that couldn't be accessed through other business strategies (Boyd, 2014).

Some businesses might take advantage of the use of partner APIs like those provided by Instagram and LinkedIn to develop their own applications. However, in the wake of the privacy concerns arising from the Cambridge Analytica incident (reported on CNBC in 2018: Meredith, 2018), Facebook-owned Instagram has imposed tighter restrictions on the use of its API by application developers. Apart from privacy issues, data owned by private companies are of a very different nature and companies are generally reluctant to open the data into the market (*Programmableweb*, 2013). On the other hand, open data made available from public agencies are usually aggregated to a level that no individ-ual identity could be revealed. It is suggested that the government should pioneer in opening up public data and set up an open API for a smart city. To achieve genuinely open data, the government needs to improve these areas.

The experiences above demonstrate that the strategy of developing SDI should be related to the characteristics and needs of the locality. The experiences of other cities' SDI projects suggest that all government services should be "digital by default," with open data and availability of APIs as the basic standard. Like Los Angeles, the government should take the lead in the development of spatial data infrastructure by lobbying various departments and the private sector to open up their geospatial data and collaborate to promote the development of various smart apps and services. A high-level government body is, therefore, required to coordinate the major tasks, including the standardization of data and setting up a framework to develop guidelines. All these will maximize the benefits of a SDI.

Generating Collective Intelligence from the Spatial Data

The second major direction is for smart cities to help citizens make better decisions. Nowadays, data collected by specific departments are generally analyzed by standard modelling tools and software, like the four-stage trip assignment model in transportation. These models are typically generated at regular intervals (typically once every year) or on an *ad hoc* basis, for example, with major new development projects. Moreover, they do not fully incorporate the human behavioral aspects and are not people-centric, as smart cities should be. Last but not least, the modelling results primarily stay within the government and are not shared with the public to facilitate their daily decision-making. Yet, these will change with smart cities. The concept of smart mobility, for example, is not just about smart transportation, that is, moving people more "smartly" in a transport system that minimizes environmental impact. The applications of ICT should allow people to make better travel decisions by providing higher accessibility to opportunities and changing people's activity pattern to a closer location previously unavailable and, hence, reduce transportation (Loo and Tsoi, 2018). Providing alternative locations for similar activities can indeed improve the choice of people on the one hand, and allow better use of resources on the other. The collective intelligence from spatial data should be able to inform citizens and enable the society to move towards different desirable sustainable transport objectives of reducing the number of trips and trip distance, better promoting public transport, encouraging active transport, cutting congestion, minimizing infrastructure expenditure (e.g., parking and expressway construction), and improving safety, among others (Loo, 2018; Loo and Tsoi, 2018). Through mining the spatial patterns of people's travel patterns, suggestions can be made not only to improve the transport system but also to change the urban form of cities with land use and other planning changes so that more sustainable mobility choices can be more easily made by the public. Nonetheless, the state-of-the-art is still very much in integrating and cleaning the spatio-temporal big data for validity and reliability (Stopher et al., 2007; Bohte and Maat, 2009; Zhou et al., 2017) and in identifying the underlying patterns in a retrospective manner (Morency et al., 2007; Ma et al., 2017) rather than in making full use of the knowledge (for example, through artificial intelligence) in directly advising individuals in making day-to-day decisions in a real-time manner. In the future, there will be much room for developing big data applications by utilizing taxi's GPS data, smart cards/ transit cards, and/or mobile phone data for use by the public to promote smart mobility (Yigitcanlar and Kamruzzaman, 2018).

Location data—the "where" factor—is especially valuable for businesses and organizations. When data are made available for the society to exchange and share through SDI, appropriate spatio-temporal analysis tools can be used to assist in scientific inquiry of pattern analysis and data mining. It is of utmost importance for policymakers and people to gain access to high quality, accurate, and timely data for their work then make well-informed smart decisions (Esri, 2017). The "Science of Where" is the science of digital transformation; the science of exploration and navigation; the science of commerce and ecology. GIS is the science of insight and innovation (Dangermond, 2017). On the one hand, a number of cities have already adopted and used location intelligence through GIS software for government officers to explore and analyze information as well as integrating the results into their daily workflow. On the other hand, allowing citizens to access the location intelligence can keep people better informed about services and business activities with regard to property, retails, logistics and transportation, taxation, entertainment, tourism, and much more (Barry, 2018).

In other words, decision-making becomes a data-driven process. A smart city fed by location intelligence, spatial analytics, and real-time data will have a wide range of applications, from optimizing supply chain management to using real-time updates for public utilities to predict demand in advance based on analytics. In a smart city with good connection and collaboration of data infrastructure, stakeholders can have quick access to the same up-to-date data on a common platform. Technologies like GIS could help enable collaboration with real-time location-based tools that allow everyone on the same platform to stay connected. For example, field workers may take their mobile devices and record the maintenance needs of the infrastructure. Data including spatio-temporal information can be instantly fed back to dashboards at the office (Barry, 2018). Location intelligence is crucial for deploying maintenance engineers during bad weather or emergency response that is both time and location dependent.

While recognizing that ICT has now been merged with traditional infrastructures, Batty (2013) argues that human capital is still fundamental to urban intelligence. With the collective intelligence generated, policymakers as well as citizens in a city can make well-informed and smart decisions with confidence with the aid of data science and analytical tools. When GIS-enabled smart maps are deployed in smart cities, there can be multifarious applications beyond map displays because the governments, businesses, and citizens can identify where the problems are, know when the problems happen, understand the surrounding context, develop appropriate applications, and make smart choices with technological tools such as open data, AI, and analytics. If the geospatial and temporal data are updated in a timely manner (Rehman et al., 2017) to be displayed on smart maps, a city can become more efficient given that citizens are able to make optimal use of the information and will be able to make smarter decisions with limited resources in the near future (Dambhare and Dambhare, 2017; Tan and Wong, 2006; Ernst and Young LLP, 2018; Barman et al., 2015).

Empowering Citizens to Plan Cities

Data transparency empowers citizens to take part in city governance. A holistic approach would encourage more public engagement, which is essential in building a smart city. After all, the ultimate model of smart city development, Smart City 3.0, should be led by citizens, not technology or the government. This is already happening in Dubai, where the data platform of Dubai Pulse under the umbrella of Smart Dubai provides an all-round city information channel, so that the public can access information of various city locations, public projects, and events. The geospatial data platform empowering the Smart Dubai initiative is a consolidation of data from different government departments followed by specialized data analysis, to form a basis for government policies. Information like the status, estimated cost, and completion time of construction works allows the government, developers, and the public to exchange ideas on urban development projects. The platform also provides a channel for the government to collect data to analyze and understand the public's opinions on the social network and respond in a timely manner.

In Singapore, one of the most well-known initiatives under SG-SPACE is the OneMap portal, which allows government departments, businesses, and individuals to create services and apps by mashing up public geospatial information via APIs to create services and apps for their specific needs. Up to now, hundreds of layers of spatial data have been created by public agencies, and a number of private projects that aim at producing new business opportunities and improving workforce productivity have been implemented.

In the United States, Los Angeles' aforementioned *GeoHub of LA City* has enabled public organizations and the community to develop mobile apps, while establishing a long-term communication channel between the public and government, stimulating innovation and creativity. Apart from enabling the collaboration between government departments and public engagement, the *GeoHub* provides data as the foundation for decision-making and legislative processes. In particular, the *GeoHub* allows the introduction of the Mayor's Dashboard, which not only helps manage the city but also allows the public to participate in and help monitor government performance. It lists the data related to people's livelihoods and divides the figures into the following four areas:

(1) Prosperity: Subdivided into three categories: economic development (such as the number of new jobs, film and TV series shooting days); economic opportunities (family rent burden, poverty—proportion of elderly and children, homeless); veterans (employment ratio and industry).
(2) Livability: Subdivided into three areas: urban services and sustainable development (such as the ratio of girls and youth participation in sports, days elapsed before potholes in streets are repaired); water and electricity (such as household daily water consumption, solar power supply); and traffic matters (e.g., traffic accidents, bus timeliness).
(3) Public security: Real-time crime information, including the type of case (violence, theft, drunkenness), and the distribution of cases on maps; the response time of police, firefighters and hotline, as well as ambulance arrival time, etc.
(4) Efficiency: Government efficiency, such as the number of civil servants, visitors to the *LA City* website, city reserves, and the non-emergency 3-1-1 hotline performance.

The Mayor's Dashboard indicates a collection of updated public concerns in Los Angeles, including the unemployment rate, the number of new jobs, the average number of days for street pothole repair, public service hotline response time, the crime rate, the reserve fund balance, etc. The *GeoHub* and its related open data policy allow Los Angeles to rank as the city in the United States with the greatest number of open data applications. The spatial platform actually supports and encourages active public participation in all areas of city life.

Monitoring and Enhancing City Performance

With reference to the main purpose(s) of smart city development defined by the local community, specific key performance indicators should be established to chart the progress towards these desirable goals. For instance, under "smart living," various indicators of renewable energy and their spatio-temporal distribution should be continuously monitored. Under "smart governance," the number of complaints and their spatio-temporal distribution needs to be monitored and actions taken to address them within specific time frame. Smart mapping should target these key indicators which are not primarily used and understood by professionals but the general public. At a theoretical level, these spatial indicators can help researchers to understand the urban hierarchy of cities within a region and a country (Wang and Loo, 2018).

Gartner, a market research company, predicted that there will be 8.4 billion objects in the world that are connected to IoT by 2017. By 2020, the number will increase by more than double to 20.4 billion. Objects connected to IoT in Greater China, the United States, and Western Europe account for nearly 70 percent of the world's total. The fields of applications include automobiles, televisions, and digital set-top boxes in the consumer market, as well as smart meters and security surveillance cameras for commercial use. Mapping-powered smart services like smart parking, smart traffic lights, digital travel, and smart grid have proven to bring significant benefits in many cities.

According to the global real-time traffic information company INRIX, drivers in New York City spend an average of 107 hours per year looking for parking (See inrix.com, accessed April 3, 2018). Meanwhile, the US Department of Transportation stated that traffic congestion costs the country US$87.2 billion from wasted fuel and loss of productivity (Federal Highway Administration, 2017). A large amount of existing public infrastructure is underutilized, or used inefficiently due to a lack of real-time information. As a result, smart traffic lights or Adaptive Signal Control Technology has been applied in some areas of the United States to allow traffic lights to adjust according to traffic conditions. This adjustment has improved travel time by more than 10 percent (Federal Highway Administration, 2017).

In China, Alibaba Cloud has also worked with the Hangzhou government to provide a smart traffic management service called Hangzhou City Brain (Ko, 2016). Its primary goal is to reduce traffic congestion through video and image recognition technologies with images from about 50,000 roadside surveillance cameras. A green light, for example, can be automatically extended when the system detects a vehicle coming, helping to shorten the waiting time. As a result, congestion has been reduced and speed of the traffic increased up to 11 percent.

The Smart Dubai initiative is also visionary in the monitoring of city performances. For example, the Smart Mobility application under the initiative collects anonymous public mobile phone data to analyze the mode of commuter traffic flow, so that officials responsible for transport planning can formulate measures to reduce traffic congestion and improve traffic safety. In addition, the interactive platform is also helping the city to reach its target of obtaining 75 percent of its energy use from renewable source by 2050. It also uses mobile applications for the public and commercial organizations to calculate the suitable location and number of solar panels to be installed. The Dubai Design District is also worth mentioning in that it is an experimental area full of IoT sensors, providing property management information such as energy consumption, temperature, and utilization which help organizations in the region save operational expenses.

In fact, Hong Kong has been deploying IoT for a time in a number of areas. A recent *Report on Smart City Blueprint for Hong Kong* (PricewaterhouseCoopers Advisory Services Limited, 2017) mentioned a variety of public infrastructures that have applied the relevant technologies in five major departments related to the six components of a smart city from smart economy, smart environment, smart government, smart living, smart mobility, and smart people. Some notable examples are the Water Supplies Department actively installing sensors in the water supply network to detect water pressure, so as to build the Water Intelligent Network to enhance maintenance efficiency and reduce water wastage due to water leakage; the Drainage Services Department using an ultrasonic sensor to detect the water level of the manhole for programming the sequence of maintenance; the Transport Department installing sensors at busy road junctions to monitor traffic conditions; the Customs Department using electronic locks (E-lock) and GPS to monitor the clearance of goods in and out of the Mainland; and the Civil Engineering and Development Department monitoring the possibility of landslide on slopes through sensors installed in retaining walls. Nonetheless, as in all the smart city initiatives, the integration of these piecemeal efforts at the city level is timely and critical. One needs to find a way to connect various existing smart city initiatives and to identify the gaps. Putting all these data on a common spatial data platform will allow a holistic and comprehensive smart city policy to be formulated, implemented, and monitored.

Conclusion

All in all, different agencies (in both public and private sectors) and individuals need a common platform where all dynamic spatio-temporal data can be shared, analyzed, and put to good use by experts and smart systems for improving people's quality of life, without requiring users to understand the complexity behind these systems. A smart system needs to be simple and easy to use. Nowadays, there are just too many components and too many different initiatives that make citizens confused about where to get information, where to voice opinions, where to get things done. The next step is to make a "one-stop" platform easy for people with different abilities to use and to make the streamlining of actions and accountability possible. So, integration is the key and the map offers incredible flexibility for all six core components of the smart city to be embedded in, from the infrastructure, to the operations and maintenance, to the need to respond to sudden disruptions (resilience). So, smart mapping is a priority that needs to be supported as the platform for advancing all six core components of the smart city and for accelerating the move in the four major directions towards the ultimate aim of creating a safe, efficient, and sustainable living environment and better quality of life.

The significance of this study lies in presenting the first systematic overview of the potential role of maps in smart cities, where IoT are constantly generating valuable real-time socio-temporal data which are not used at all or not fully utilized by major stakeholders, including academics, policymakers, professionals, and the public, to achieve smart city goals. In this paper, the major challenges, directions of change, and implications are discussed with respect to data collection, data analysis, institutional set-up, and users of data, in relation to the six domains of the smart city wheel. A clear limitation of this study is that it only reflects one possible conceptualization of the relationship between maps and smart cities. In the future, more in-depth research which captures the costs and real value (beyond economic benefits) of specific smart mapping initiatives in relation to the four research directions will be needed for evaluating smart mapping as a key dimension of a smart city strategy.

Acknowledgments

This research was supported by the National Natural Science Foundation of China [grant no. 41671154].

ORCID

Becky P.Y. Loo ⓘ http://orcid.org/0000-0003-0822-5354

References

V. Albino, B. Umberto, and R.M. Dangelico, "Smart Cities: Definitions, Dimensions, Performance, and Initiatives," *Journal of Urban Technology* 22: 1 (2015) 3–21.

M. Angelidou, . Angelidou,rban Technologyimensions, Pe," *Cities* 41: Supp 1 (2014) S3–S11.

K. Ashton, "That 'Internet of Things' Thing: In the Real World, Things Matter More than Ideas,ni*RFID Journal* (June 22, 2009) <http://www.rfidjournal.com/articles/view?4986> Accessed April 3, 2018.

S. Barman, A.K. Barman, B. Panda, and P. Sharma, "Implementation of a Smart Map Using Spatial Oracle," paper presented at the Third International Conference on Computer, Communication, Control and Information Technology (Hooghly, February 7–8, 2015).

J. Barry, "3 Steps to Make GIS Part of a Successful Smart City," *Smartcitiesdive* (June 8, 2018) <https://www.smartcitiesdive.com/news/3-steps-to-make-gis-part-of-a-successful-smart-city/525245/> Accessed April 3, 2018.

M. Batty, *The New Science of Cities* (Cambridge: The MIT Press, 2013).

M. Batty, A. Hudson-Smith, R. Milton, and A. Crooks, "mith, R. Milton, and A. Crooks, cial Topics," *Annals of GIS* 16 (2010) 1–13.

N. Baucells, C. Moreno, and R.M. Arce, "Mapping Smart Cities Situation plus CITIES: The Spanish Case," paper presented at the First International Conference on Smart Cities (Malaga, June 15–17, 2016).

W. Bohte and K. Maat, "Deriving and Validating Trip Purposes and Travel Modes for Multi-day GPS-based Travel Surveys: A Large-scale Application in the Netherlands," *Transportation Research Part C: Emerging Technologies* 17: 3 (2009) 285–297.

M. Boyd, "Private, Partner or Public: Which API Strategy is Best for Business?" *ProgrammableWeb* (21 February 2014) <https://www.programmableweb.com/news/private-partner-or-public-which-api-strategy-best-business/2014/02/21> Accessed April 3, 2018.

M.A. Celdran-Bernabeu, J.-N. Mazon, J.A. Ivars-Baidal, and J.F. Vera-Rebollo, ". Vera-Rebollo, tner-or-public-which-api-str," *Cuadernos de Turismo* 41 (2018) 655–658.

T.Y. Ching and J. Ferreira, "Smart Cities: Concepts, Perceptions and Lessons for Planners," paper presented at Computers in Urban Planning and Urban Management (Boston, July 7–10, 2015).

J. Chow, Building a Smart City: How Hong Kong Can Improve its Spatial Data Infrastructure," *South China Morning Post*, (March 17, 2017). <https://www.scmp.com/special-reports/property/topics/weekend-property/article/2079503/building-smart-city-how-hong-kong> Accessed February 19, 2019.

Cisco, *Smart City Framework: A Systematic Process for Enabling Smart+Connected Communities* (San Jose: Cisco, 2012).

B. Cohen, "The Smart City Wheel," Smart Circle (May 14, 2013) <https://www.smart-circle.org/smartcity/blog/boyd-cohen-the-smart-city-wheel/> Accessed April 3, 2018.

Dalberg Global Development Advisors, *Smart Maps for Smart Cities: Urban India's Billion+ Opportunity* (n.p.: Dalberg, 2015).

A.U. Dambhare and A.F. Dambhare "Smart Map for Smart City," paper presented at International Conference on Innovative Mechanisms for Industry (Bangalore, February 21–23, 2017).

J. Dangermond, "Understanding the Science of Where," *ArcGIS Blog* (October 6, 2017) <https://www.esri.com/arcgis-blog/products/product/uncategorized/understanding-the-science-of-whe re/> Accessed April 3, 2018.

Deloitte, "Smart Cities: The importance of a smart ICT infrastructure for smart cities," *Stokab* (January, 2017) <https://www.stokab.se/Documents/Nyheter%20bilagor/SmartCityInfraEn.pdf> Accessed April 3, 2018.

Ernst and Young LLP, Start Your Journey with the Supply Chain Smart Maps, *EY* (2018) <https://www.ey.com/Publication/vwLUAssets/EY-start-your-journey-with-the-supply-chain-smart-map s/$File/EY-start-your-journey-with-the-supply-chain-smart-maps.pdf> Accessed April 3, 2018.

Esri, *GIS Best Practices: Spatial Data Infrastructure* (2010) <https://www.esri.com/library/bestpractices/spatial-data-infrastructure.pdf> Accessed February 19, 2019.

Esri, "Los Angeles Launched GeoHub" (Spring, 2016) <https://www.esri.com/esri-news/arcnews/spring16articles/los-angeles-launched-geohub> Accessed 19 February, 2019.

Esri, "Esri and The Science of Where, Supporting and Incorporating Science," *Esri .<com* (Spring, 2017) <https://www.esri.com/~/media/Files/Pdfs/news/arcuser/0517/esri-and-the-science-of-w here.pdf> Accessed April 3, 2018.

Esri Malaysia, "Smart Maps Transforming Crime Fighting," *Esri Malaysia* (July 13, 2015) <http://esrimalaysia.com.my/news/smart-maps-transforming-crime-fighting-says-global-expert-nar-43> Accessed April 3, 2018.

N. Gershenfeld, R. Krikorian, and D. Cohen "The Internet of Things," *Scientific America* 291: 4 (2004) 76–81.

R. Giffinger, H. Karmar, and G. Haindl, "The Role of Rankings in Growing City Competition," paper presented at the Xi Eura Conference (Milan, October 9–11, 2008).

J.R. Gil-Garcia, T.A. Pardo, and T. Nam, "What Makes a City Smart? Identifying Core Components and Proposing an Integrative and Comprehensive Conceptualization," *Information Polity* 20 (2015) 61–87.

J. Ginsberg, M.H. Mohebbi, R.S. Patel, L. Brammer, M.S. Smolinski, and L. Brilliant, "and L. Brilliant, r, M. Identifying Core Components and Prop," *Nature* 457 (2009) 1012–1014.

M.F. Goodchild, "Citizens as Sensors: The World of Volunteered Geography," *GeoJournal* 69 (2007) 211–221.

M.F. Goodchild, "Looking Forward Again: Four Thoughts on the Future of GIS in 2015 and Beyond," *Esri ArcWatch* (February, 2015) <http://www.esri.com/esri-news/arcwatch/0215/four-thoughts-on-the-future-of-gis-in-2015-and-beyond> Accessed April 3, 2018.

M. Graham, "Geography/internet: Ethereal Alternate Dimensions of Cyberspace or Grounded Augmented Realities?" *The Geographical Journal* 179 (2013)177–182.

S. Hahmann, D. Burghardt, and B. Weber, "and B. Weber, l Journalreal Alternate Dimensions "??? Towards a Research Framework: Using the Semantic Web for (In) Validating this Famous Geo Assertion," paper presented at AGILE 2011(Salt Lake City, August 8–12, 2011).

L.L. Hill, Georeferencing: The Geographic Associations of Information (Cambridge: MIT Press, 2009).

M.A. Hossen, "Participatory Mapping for Community Empowerment," *Asian Geographer* 33: 2 (2016) 97–113.

IBM, *A Vision of Smarter Cities*, (New York: IBM Institute of Business Value, 2009).

K. Kim, "Interacting Socially with the Internet of Things (IoT): Effects of Source Attribution and Specialization in Human–IoT Interaction," *Journal of Computer-Mediated Communication* 21:6 (2016) 420–435.

C. Ko, ibaba Cloud Bridges Worldphy Opportunities, Challenges and Risk*ComputerWorldHK* (October 13, 2016) <http://cw.com.hk/news/alibaba-cloud-bridges-worlds-longest-distance-city-brain?page = 0,1> Accessed April 3, 2018.

W.W.Y. Lam, S. Rasouli, and H.J.P. Timmermans, kSpecial Issue on Advances in Spatiotemporal Transport Analysis,pe*Travel Behaviour and Society* 1:1 (2014).

I. Lee and K. Lee, "The Internet of Things (IoT): Applications, Investments, and Challenges for Enterprises," *Business Horizons* 58:4 (2015) 431–440.

C. Li, P. Liu, J. Yin, and X. Liu, "The Concept, Key Technologies and Applications of Temporal–spatial Information Infrastructure," *Geo-spatial Information Science* 19: 2 (2016) 148–156.

S.Y. Loh and V. Khoo, "Spatially Enabled Singapore through Singapore Geospatial Collaborative Environment (SG–SPACE)," in A. Rajabifard, J. Crompvoets, M. Kalantari, and B. Kok, eds, Spatially Enabling Society: Research, Emerging Trends and Critical Assessment (Leuven: Leuven University Press, 2010) 111–116.

B.P.Y. Loo, *E-Society* (New York: Nova Science, 2012).

B.P.Y. Loo, "Strategies of Internet Development in the Asia–Pacific Region," *Journal of Urban Technology* 14: 1 (2007) 3–22.

B.P.Y. Loo, *Unsustainable Transport and Transition in China* (London: Routledge, 2018).

B.P.Y. Loo and K.H. Tsoi, "The Sustainable Transport Pathway: A Holistic Strategy of Five Transformations," *Journal of Transport and Land Use* 11: 1 (2018) 961–980.

B.P.Y. Loo and A.Y.P. Wong, "Internet Development in Asia–Pacific: Spatial Patterns and Underlying Locational Factors," *Networks and Communication Studies* 16: 3/4 (2002) 113–134.

X. Ma, C. Liu, H. Wen, Y. Wang, and Y.J. Wu, "Understanding Commuting Patterns using Transit Smart Card Data," *Journal of Transport Geography* 58 (2017) 135–145.

S. Meredith, "Facebook-Cambridge Analytica: A Timeline of the Data Hijacking Scandal," *CNBC* (April 10, 2018) <https://www.cnbc.com/2018/04/10/facebook-cambridge-analytica-a-timeline-of-the-data-hijacking-scandal.html> Accessed April 3, 2018.

A. Metz, "How Maps Will Make People and their Cities Smarter," *Here* (September 17,2014) <https://360.here.com/2014/09/17/smart-citizens-smart-cities/> Accessed April 3, 2018.

C. Morency, M. Trépanier, and B. Agard, "Measuring Transit Use Variability with Smart-Card Data," *Transport Policy* 14: 3 (2007) 193–203.

K.U. Pfeiffer and K.B. Stevens, "Spatial and Temporal Epidemiological Analysis in the Big Data Era," *Preventive Veterinary Medicine* 122 (2015) 213–220.

PricewaterhouseCoopers Advisory Services Limited, *Report of Consultancy Study on Smart City Blueprint for Hong Kong.* (Hong Kong: PricewaterhouseCoopers Advisory Services Limited, 2017) < https://www.smartcity.gov.hk/report/full/> Accessed April 3, 2018.

ProgrammableWeb, "5 Reasons Your API is Still Private," *ProgrammableWeb* (November 18, 2013) <https://www.programmableweb.com/news/5-reasons-your-api-still-private/analysis/2013/11/18> Accessed April 3, 2018.

A. Ramaprasad, A. Sanchez-Ortiz, and T. Syn, "A Unified Definition of a Smart City," paper presented at the 16th IFIP WG 8.5 International Conference, EGOV (St Petersburg, September 4–7, 2017).

F. Rehman, I. Afyouni, A. Lbath, and S. Basalamah, "Towards Building Smart Maps from Heterogeneous Data Sources," paper presented at the 9th International Conference on Advanced Geographic Information systems, Applications and Services (Nice, 19–23 March 2017).

M. Robinson, "A History of Spatial Data Coordination," *Federal Geographic Data Committee* (May, 2008) <https://www.fgdc.gov/ngac/a-history-of-spatial-data-coordination.pdf> Accessed April 3, 2018.

B. Sadiq, F.U. Rehman, A. Ahmad, M.A. Rahman, S. Ghani, A. Murad, S. Basalamah, and A. Lbath, "A Spatio–temporal Multimedia Big Data Framework for a Large Crowd," paper presented at 2015 IEEE International Conference on Big Data (Santa Clara, 29 October–1 November 2015).

J. Steenbruggen, E. Tranos and P. Nijkamp, Steenbruggen, E. Tranos and P. Nijkamp, m-api/" graphy,Big D" *Telecommunications Policy* 39 (2015) 335–346.

D. Steudler and A. Rajabifard, Spatially Enabled Society (Copenhagen, Denmark: The International Federation of Surveyors and the Global Spatial Data Infrastructure Association, 2012) <https://www.fig.net/resources/publications/figpub/pub58/figpub58.pdf> Accessed April 3, 2018.

P. Stopher, C. FitzGerald, and M. Xu, "Assessing the Accuracy of the Sydney Household Travel Survey with GPS," *Transportation* 34: 6 (2007) 723–741.

S.C. Tan and O. Wong, "Location Aware Applications for Smart Cities with Google Maps and GIS Tools," paper presented at the 4th International Conference on Active Media Technology (Queensland, Australia, June 7–9, 2006).

A. Tascikaraoglu, "Evaluation of Spatio-temporal Forecasting Methods in Various Smart City Applications," *Renewable and Sustainable Energy Reviews* 82 (2017) 424–435.

The Masterbuilder, "Smart Maps: The Backbone for Smart Cities," *Masterbuilder* (June 22, 2015) <https://www.masterbuilder.co.in/smart-maps-the-backbone-for-smart-cities> Accessed April 3, 2018.

The White House, *Circular No. A-16* Revised (August 2002) <https://obamawhitehouse.archives.gov/omb/circulars_a016_rev/> Accessed February 19, 2019.

R.F. Tomlinson, mA Geographic Information System for Regional Planning," *Data Handling and Interpretation* (1968) 200–210.

Federal Highway Administration, *Adaptive Signal Control Technology* (US Department of Transportation, 2017) < https://www.fhwa.dot.gov/innovation/everydaycounts/edc-1/asct.cfm> Accessed April 3, 2018.

B. Wang and B.P.Y. Loo, "The Hierarchy of Cities in Internet News Media and Internet Search: Some Insights from China," *Cities* (Online First): (2018).

N. Weinmann, "Avoid Traffic Congestion by Using Real-time Information," *Esri* (October 15, 2014) < https://community.esri.com/docs/DOC-2576> Accessed April 3, 2018.

T. Yigitcanlar and M. Kamruzzaman, "Smart Cities and Mobility: Does the Smartness of Australian Cities Lead To Sustainable Commuting Patterns?" *Journal of Urban Technology* 26:2 (2019).

J.J. Zhao, C. Tian, F. Zhang, C.Z. Xu and S.Z. Feng, "Understanding Temporal and Spatial Travel Patterns of Individual Passengers by Mining Smart Card Data," paper presented at the 17th International Conference on Intelligent Transportation, (Qingdao, October 8–11, 2014).

M. Zhou, D. Wang, Q. Li, Y. Yue, W. Tu, and R. Cao, "Impacts of Weather on Public Transport Ridership: Results from Mining Data from Different Sources," *Transportation Research Part C: Emerging Technologies*75:(2017) 17–29.

Towards Post-Anthropocentric Cities: Reconceptualizing Smart Cities to Evade Urban Ecocide

Tan Yigitcanlar ⓘ, Marcus Foth ⓘ, and Md. Kamruzzaman ⓘ

ABSTRACT
This short piece acts as a coda to this journal's special issue on "Smart Cities and Innovative Technologies." First, it provides a retrospective view of the origins of the smart city concept. The paper, secondly, presents the most recent perspectives on the new interpretations of the smart city notion. It then provides a commentary on the potential directions for a better reconceptualization of smart cities to evade a most likely urban ecocide. Lastly, the short communication concludes by asking two critical questions: (a) Will urban scholars, planners, designers, and activists be able to convince urban policymakers and the general public of the need for a post-anthropocentric urban turnaround? (b) How do the public, private, and academic sectors along with communities pave the way for post-anthropocentric cities and more-than-human futures?

Introduction: Can Technology Save Us?

The current Anthropocene era is characterized by greenhouse gas emissions and human domination (Crutzen and Steffen, 2003). As a result, the world is being confronted with severe environmental, economic, and social crises (Moore, 2017). This is combined with rapid urbanization, increased mobilization, heightened globalization, ruthless neoliberal capitalism, vigorous industrialization, intensified agriculture, excessive consumption, and highly materialized lifestyles (Yigitcanlar and Dizdaroglu, 2015; Monbiot, 2016). At this dire strait, contemporary urban policy and practice tend to place all their bets on technology as a panacea to ensure our survival (Wiig, 2015). Yet, can technology alone really save us?

Rapid advancements on the technology front—particularly as a result of the second wave of the digital revolution and the fourth industrial revolution—along with aggressive marketing by technology companies gave policymakers and urban administrators a false hope (Söderström et al., 2014). The hope was that the effects of global-scale environmental and socioeconomic crises could be reversed through feasible technology solutions. Consequently, the amalgamation of technology and the city is widely seen as an effective instrument to manage the challenges that cities and societies are facing (Yigitcanlar, 2016). This fusion of technology and the city is referred to as "smart cities," a concept that has evolved through different stages (Foth, 2018; Yigitcanlar et al., 2018).

The First Generation: Intelligent Cities

Even though the smart city concept was popularized by the technology companies around the mid-2000s, its origin dates back to the intelligent city notion of the 1990s. The "intelligent cities" paradigm brought together the trajectories of the knowledge and innovation economy, and the spread of the Internet and World Wide Web as major technological innovations (Komninos, 2011). Intelligent cities (the first-generation smart city) were the realm of technology companies providing innovative technologies to local governments in order to improve and optimize the efficiency of specific city functions. This conceptualization was heavily expert-focused and almost no opportunity was given for citizens to participate in the decision-making process.

The Second Generation: Smart Cities

In the late-2000s, as an extension of the intelligent city movement, the "smart cities" concept emphasized a greater degree of involvement of local authorities in deploying smart technologies (Yigitcanlar, 2015). Targeting city infrastructure and services, these technologies established a new digital data layer to drive efficiencies through smart meters and shared mobility. This second-generation smart city employs sensors and other Internet-of-Things (IoT) devices with a growing emphasis on urban informatics, urban science, and data analytics aimed at solving urban problems (Lim and Taeihagh, 2018). Yet, the highly top-down approach in investment and governance remains—leaving only limited room for the community's voice in the policymaking process.

The Third Generation: Responsive Cities

As a reaction to the conceptualization and practice limitations of smart cities, a new type of city model is envisaged: A city that provides citizens with active engagement in and usage of smart solutions to improve living standards and urban sustainability. This is referred to as "responsive cities" (Goldsmith and Crawford, 2014). These cities restore the citizen's right to the digital city by giving citizens power to use smart technology to contribute to planning, design, and management of their cities (Foth et al., 2015). The responsive city (the third-generation smart city) relies on IoT and mobile devices communicating autonomously with the aim of improving urban life.

The Challenge: Can Smart Cities Address the Causes of Our Urban Ills?

The progression from intelligent to smart and from there to responsive cities consisted positive moves and contributed cumulatively to the urban policymaking practice. However, city innovation remains largely technocentric with much needed governance, policy, and regulatory reform lagging behind in both speed and scope (Noy and Givoni, 2018). Technocratic approaches generate serious doubts about their capability of addressing the aforementioned root problems causing environmental, economic, and social crises (Kunzmann, 2014).

In recent years, various international, national, and regional city ranking exercises listed the best performing smart cities, and various studies provided insights into smart city best practices (Giffinger and Gudrun, 2010). These exercises and studies celebrated the achievements of a number of global smart cities—including Amsterdam, Barcelona, Boston, London, New York, Paris, San Francisco, Seoul, Singapore, Stockholm, Tokyo, and Vienna. However, a closer look into the environmental performance of these cities reveals unsustainable levels of per capita greenhouse gas emissions despite some regulations (Hoornweg et al., 2011; Arbolino et al., 2017).

Moreover, recent empirical studies have reported that smart cities are not after all that smart as they fail to live up to sustainability expectations. For example, a recent study of 15 UK smart cities found no evidence that urban smartness contributes to sustainable outcomes (Yigitcanlar and Kamruzzaman, 2018a). Other research on Australian cities revealed the smartness of cities does not lead to sustainable commuting patterns (Yigitcanlar and Kamruzzaman, 2018b). Additionally, studies on smart cities in Africa and South Korea—including Songdo, recognized as the world's "smartest" city—evidenced the environmental downfalls of these ambitious projects (Watson, 2014; Yigitcanlar and Lee, 2014). Furthermore, it is argued that cities cannot be truly smart unless they produce zero waste (Zaman and Lehmann, 2013) and make a net positive contribution to the ecosystem (Birkeland, 2012).

While useful to describe the changing attitude of local governments towards smart city investments, the trend from "intelligent" to "smart" to "responsive" cities remains highly constrained by the focus on technology and technical systems (Anthopoulos, 2017). This in turn begs questions about the depletion of rare earth metals and the accumulation of e-waste. A technocratic approach is also not adequate in recognizing our ecological entanglements with nature (Houston et al., 2018). It does not avoid the ecocide and existential crisis we face in light of forthcoming catastrophes of the Anthropocene era (MacDougall et al., 2013)—such as the ecosystem collapse of the Great Barrier Reef (Pandolfi et al., 2003).

The Fourth Generation: What Does a Truly Smart and Sustainable City Look Like?

Current smart city practice is generating a Frankenstein urbanism by forcing the union of different and incompatible elements in cities—in a disingenuous attempt of addressing quality of life and sustainability (Cugurullo, 2018). There is, hence, an urgency to reconceptualize urban planning, design, and development paradigms and act accordingly and immediately. In such reconceptualizations that question human exceptionalism (Houston et al., 2018), urban space cannot be seen as an entity separate from nature and thus it cannot be designed just or primarily for humans. Decentering the human in urban design (Forlano, 2016) will help to develop post-anthropocentric cities or more-than-human cities (the fourth-generation smart city?) that are truly smart, sustainable, and equitable (Foth, 2017; Franklin, 2017).

Concluding Remarks: Towards a Post-Anthropocentric Urban Turnaround?

The current smart city practice, at its best, is a zero-sum game for sustainability—environmental gains are cancelled out by the impact of increased technology and energy use (Ahvenniemi et al., 2017). The biggest challenge now is finding a way to change our mentality and politics on how we shape our cities, societies, and the environment. We need to move forward instantaneously and quickly by focusing on an ecological human settlement theory (Liaros, 2018) that will create cohabitation spaces to house humans and non-humans in a sustainable and inclusive way in the post-anthropocentric cities of tomorrow.

The sixth extinction is already upon us (Celabllos et al., 2015). Building post-anthropo-centric cities for more-than-human futures might be the last resort for humankind to evolve and avoid extinction in the not too distant future. Nevertheless, in this instance, human civilization is standing at the crossroads. A number of critical decisions must be taken and implemented immediately—for example, there must be a moving away from aggressive population increases, and from viewpoints where urban expansion and econ-omic growth are dominant. Furthermore, the right answers to the following questions will also be extremely critical to our future existence on the planet and its living conditions:

(1) Will urban scholars, planners, designers, and activists be able to convince urban pol-icymakers and the general public of the urgent need for a post-anthropocentric urban turnaround?

(2) How can we—jointly with public, private, and academic sectors along with commu-nities—pave the way for post-anthropocentric cities and more-than-human futures?

Disclosure Statement

No potential conflict of interest was reported by the authors.

ORCID

Tan Yigitcanlar ⓘ http://orcid.org/0000-0001-7262-7118
Marcus Foth ⓘ http://orcid.org/0000-0001-9892-0208
Md. Kamruzzaman ⓘ http://orcid.org/0000-0001-7113-942X

Bibliography

H. Ahvenniemi, A. Huovila, I. Pinto-Seppä, and M. Airaksinen, "What are the Differences Between Sustainable and Smart Cities?" *Cities* 60 (2017) 234–245.

L. Anthopoulos, "Smart Utopia VS Smart Reality: Learning by Experience from 10 Smart City Cases," *Cities* 63 (2017) 128–148.

R. Arbolino, F. Carlucci, A. Cirà, G. Ioppolo, and T. Yigitcanlar, "Efficiency of the EU Regulation on Greenhouse Gas Emissions in Italy: The Hierarchical Cluster Analysis Approach," *Ecological Indicators* 81 (2017) 115–123.

J. Birkeland, *Design for Sustainability: A Sourcebook of Integrated Ecological Solutions* (Oxford: Routledge, 2012).

G. Ceballos, P. Ehrlich, A. Barnosky, A. García, R. Pringle, and T. Palmer, "Accelerated Modern Human-Induced Species Losses: Entering the Sixth Mass Extinction," *Science Advances* 1 (2015) e1400253.

P. Crutzen and W. Steffen, "How Long Have We Been in the Anthropocene Era?" *Climatic Change* 61 (2003) 251–257.

F. Cugurullo, "Exposing Smart Cities and Eco-Cities: Frankenstein Urbanism and the Sustainability Challenges of the Experimental City," *Environment and Planning A* 50 (2018) 73–92.

L. Forlano, "Decentering the Human in the Design of Collaborative Cities," *Design Issues* 32 (2016) 42–54.

M. Foth, "The Next Urban Paradigm: Cohabitation in the Smart City," *IT-Information Technology* 59 (2017) 259–262.

M. Foth, "Participatory Urban Informatics: Towards Citizen-Ability," *Smart and Sustainable Built Environment* 7 (2018) 4–19.

M. Foth, M. Brynskov, and T. Ojala, *Citizen's Right to the Digital City: Urban Interfaces, Activism, and Placemaking* (Singapore: Springer, 2015).

A. Franklin, "The More-Than-Human City," *The Sociological Review* 65 (2017) 202–217.

S. Goldsmith and S. Crawford, *The Responsive City: Engaging Communities Through Data-Smart Governance* (London: John Wiley & Sons, 2014).

R. Giffinger and H. Gudrun, "Smart Cities Ranking: An Effective Instrument for the Positioning of the Cities?" *Architecture, City and Environment* 4 (2010) 7–26.

D. Hoornweg, L. Sugar, and C. Trejos-Gomez, "Cities and Greenhouse Gas Emissions: Moving Forward," *Environment and Urbanization* 23 (2011) 207–227.

D. Houston, J. Hillier, D. MacCallum, W. Steele, and J. Byrne, "Make Kin, Not Cities! Multispecies Entanglements and 'Becoming-World' in Planning Theory," *Planning Theory* 17 (2018) 190–212.

N. Komninos, "Intelligent Cities: Variable Geometries of Spatial Intelligence," *Intelligent Buildings International* 3 (2011) 172–188.

K. Kunzmann, "Smart Cities: A New Paradigm of Urban Development," *Crios* 1 (2014) 9–20.

S. Liaros, "An Ecological Human Settlement Theory," *GreenAgenda.org* (June 14, 2018) <https://greenagenda.org.au/2018/06/ecological-human-settlement-theory> accessed July 30, 2018.

H. Lim and A. Taeihagh, "Autonomous Vehicles for Smart and Sustainable Cities: An In-Depth Exploration of Privacy and Cybersecurity Implications," *Energies* 11 (2018) 1062.

A. MacDougall, K. McCann, G. Gellner, and R. Turkington, "Diversity Loss with Persistent Human Disturbance Increases Vulnerability to Ecosystem Collapse," *Nature* 494 (2013) 86–89.

G. Monbiot, *How Did We Get Into This Mess? Politics, Equality, Nature* (London: Verso Books, 2016).

J. Moore, "The Capitalocene, Part I: On the Nature and Origins of our Ecological Crisis," *The Journal of Peasant Studies* 44 (2017) 594–630.

K. Noy and M. Givoni, "Is 'Smart Mobility' Sustainable? Examining the Views and Beliefs of Transport's Technological Entrepreneurs," *Sustainability* 10 (2018) 422.

J. Pandolfi, R. Bradbury, E. Sala, T. Hughes, K. Bjorndal, R. Cooke, and R. Warner, "Global Trajectories of the Long-Term Decline of Coral Reef Ecosystems," *Science* 301 (2003) 955–958.

O. Söderström, T. Paasche, and F. Klauser, "Smart Cities as Corporate Storytelling," *City* 18 (2014) 307–320.

V. Watson, "African Urban Fantasies: Dreams or Nightmares?" *Environment and Urbanization* 26 (2014) 215–231.

A. Wiig, "IBM's Smart City as Techno-Utopian Policy Mobility," *City* 19 (2015) 258–273.

T. Yigitcanlar, "Smart Cities: An Effective Urban Development and Management Model?" *Australian Planner* 52 (2015) 27–34.

T. Yigitcanlar, *Technology and the City: Systems, Applications and Implications* (New York: Routledge, 2016).

T. Yigitcanlar and D. Dizdaroglu, "Ecological Approaches in Planning for Sustainable Cities: A Review of the Literature," *Global Journal of Environmental Science and Management* 1 (2015) 159–188.

T. Yigitcanlar and M. Kamruzzaman, "Does Smart City Policy Lead to Sustainability of Cities?" *Land Use Policy* 73 (2018a) 49–58.

T. Yigitcanlar and M. Kamruzzaman, "Smart Cities and Mobility: Does the Smartness of Australian Cities Lead to Sustainable Commuting Patterns?" *Journal of Urban Technology* (2018b) https://doi.org/10.1080/10630732.2018.1476794.

T. Yigitcanlar, M. Kamruzzaman, L. Buys, G. Ioppolo, J. Sabatini-Marques, E. Costa and J. Yun, "Understanding Smart Cities: Intertwining Development Drivers with Desired Outcomes in a Multidimensional Framework," *Cities* 81 (2018) 145–160.

T. Yigitcanlar and S. Lee, "Korean Ubiquitous-Eco-City: A Smart-Sustainable Urban Form or a Branding Hoax?" *Technological Forecasting and Social Change* 89 (2014) 100–114.

A. Zaman and S. Lehmann, "The Zero Waste Index: A Performance Measurement Tool for Waste Management Systems in a Zero Waste City," *Journal of Cleaner Production* 50 (2013) 123–132.

Index

Page numbers in **bold** refer to tables and those in *italic* refer to figures.